Selected Papers
in *K*-Theory

Recent Titles in This Series

- 154 V. A. Artamonov, et al., Selected Papers in K-Theory
- 153 S. G. Gindikin, Editor, Singularity Theory and Some Problems of Functional Analysis
- 152 H. Draškovičová, et al., Ordered Sets and Lattices II
- 151 I. A. Aleksandrov, L. A. Bokut', and Yu. G. Reshetnyak, Editors, Second Siberian Winter School "Algebra and Analysis"
- 150 S. G. Gindikin, Editor, Spectral Theory of Operators
- 149 V. S. Afraĭmovich, et al., Thirteen Papers in Algebra, Functional Analysis, Topology, and Probability, Translated from the Russian
- 148 A. D. Aleksandrov, O. V. Belegradek, L. A. Bokut', and Yu. L. Ershov, Editors, First Siberian Winter School in Algebra and Analysis
- 147 I. G. Bashmakova, et al., Nine Papers from the International Congress of Mathematicians 1986
- 146 L. A. Aĭzenberg, et al., Fifteen Papers in Complex Analysis
- 145 S. G. Dalalyan, et al., Eight Papers Translated from the Russian
- 144 S. D. Berman, et al., Thirteen Papers Translated from the Russian
- 143 V. A. Belonogov, et al., Eight Papers Translated from the Russian
- 142 M. B. Abalovich, et al., Ten Papers Translated from the Russian
- 141 Kh. Drashkovicheva, et al., Ordered Sets and Lattices
- 140 V. I. Bernik, et al., Eleven Papers Translated from the Russian
- 139 A. Ya. Aĭzenshtat, et al., Nineteen Papers on Algebraic Semigroups
- 138 I. V. Kovalishina and V. P. Potapov, Seven Papers Translated from the Russian
- 137 V. I. Arnol'd, et al., Fourteen Papers Translated from the Russian
- 136 L. A. Aksent'ev, et al., Fourteen Papers Translated from the Russian
- 135 S. N. Artemov, et al., Six Papers in Logic
- 134 A. Ya. Aĭzenshtat, et al., Fourteen Papers Translated from the Russian
- 133 R. R. Suncheleev, et al., Thirteen Papers in Analysis
- 132 I. G. Dmitriev, et al., Thirteen Papers in Algebra
- 131 V. A. Zmorovich, et al., Ten Papers in Analysis
- 130 M. M. Lavrent'ev, et al., One-dimensional Inverse Problems of Mathematical Physics
- 129 S. Ya. Khavinson; translated by D. Khavinson, Two Papers on Extremal Problems in Complex Analysis
- 128 I. K. Zhuk, et al., Thirteen Papers in Algebra and Number Theory
- 127 P. L. Shabalin, et al., Eleven Papers in Analysis
- 126 S. A. Akhmedov, et al., Eleven Papers on Differential Equations
- 125 D. V. Anosov, et al., Seven Papers in Applied Mathematics
- 124 B. P. Allakhverdiev, et al., Fifteen Papers on Functional Analysis
- 123 V. G. Maz'ya, et al., Elliptic Boundary Value Problems
- 122 N. U. Arakelyan, et al., Ten Papers on Complex Analysis
- 121 D. L. Johnson, The Kourovka Notebook: Unsolved Problems in Group Theory
- 120 M. G. Kreĭn and V. A. Jakubovič, Four Papers on Ordinary Differential Equations
- 119 V. A. Dem'janenko, et al., Twelve Papers in Algebra
- 118 Ju. V. Egorov, et al., Sixteen Papers on Differential Equations
- 117 S. V. Bočkarev, et al., Eight Lectures Delivered at the International Congress of Mathematicians in Helsinki, 1978
- 116 A. G. Kušnirenko, A. B. Katok, and V. M. Alekseev, Three Papers on Dynamical Systems
- 115 I. S. Belov, et al., Twelve Papers in Analysis

(Continued in the back of this publication)

American Mathematical Society

TRANSLATIONS

Series 2 • Volume 154

Selected Papers in *K*-Theory

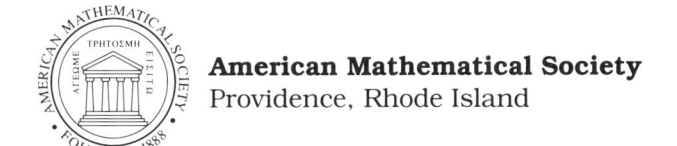

American Mathematical Society
Providence, Rhode Island

1991 *Mathematics Subject Classification.* Primary 11R, 11S, 13C, 16A, 16H, 16S, 18F, 19A, 19B, 19D, 19E, 19K, 20C, 20G, 46J, 46L, 47B, 47D, 55N; Secondary 11R, 12E, 16A, 16E, 18E, 20E, 20F, 46A, 47A.

Library of Congress Cataloging-in-Publication Data
Selected papers in K-theory.
   p. cm. – (American Mathematical Society translations; ser. 2, v. 154)
   Includes bibliographical references.
   ISBN 0-8218-7504-3 (acid free paper)
   1. K-theory. I. American Mathematical Society. II. Series.
QA3.A572   ser. 2, vol. 154
[QA612.33]
510 s—dc20   92-18202
[514'.23]   CIP

    COPYING AND REPRINTING. Individual readers of this publication, and nonprofit libraries acting for them, are permitted to make fair use of the material, such as to copy an article for use in teaching or research. Permission is granted to quote brief passages from this publication in reviews, provided the customary acknowledgment of the source is given.
    Republication, systematic copying, or multiple reproduction of any material in this publication (including abstracts) is permitted only under license from the American Mathematical Society. Requests for such permission should be addressed to the Manager of Editorial Services, American Mathematical Society, P.O. Box 6248, Providence, Rhode Island 02940-6248.
    The appearance of the code on the first page of an article in this book indicates the copyright owner's consent for copying beyond that permitted by Sections 107 or 108 of the U.S. Copyright Law, provided that the fee of $1.00 plus $.25 per page for each copy be paid directly to the Copyright Clearance Center, Inc., 27 Congress Street, Salem, Massachusetts 01970. This consent does not extend to other kinds of copying, such as copying for general distribution, for advertising or promotional purposes, for creating new collective works, or for resale.

            Copyright ©1992 by the American Mathematical Society. All rights reserved.
                       Printed in the United States of America.
                The American Mathematical Society retains all rights
                    except those granted to the United States Government.
       The paper used in this book is acid-free and falls within the guidelines
                established to ensure permanence and durability. ∞
                  This publication was typeset using $\mathcal{A}_{\mathcal{M}}\mathcal{S}$-TEX,
             the American Mathematical Society's TEX macro system.
                 10 9 8 7 6 5 4 3 2 1   97 96 95 94 93 92

Contents

V. P. Platonov and A. A. Sharomet, On the congruence problem for linear groups over arithmetic rings — 1

V. P. Platonov, Birational properties of the reduced Whitehead group — 7

V. A. Artamonov, Projective modules over group rings of nilpotent groups — 11

S. V. Vostokov and I. B. Fesenko, On torsion in higher Milnor functors for multidimensional local fields — 25

A. A. Klyachko, K-theory of Demazure models — 37

V. A. Lipnitskiĭ, Norms in fields of real algebraic functions, and the reduced Whitehead group — 47

V. A. Lipnitskiĭ, The reduced Whitehead group of weakly and totally ramified division algebras — 53

V. I. Kaskevich, The structure of unitary and orthogonal groups over Henselian discretely normed Azumaya algebras — 57

V. I. Yanchevskiĭ, Reduced unitary Whitehead groups and noncommutative rational functions — 63

I. I. Voronovich and V. I. Yanchevskiĭ, Rational splitting fields of simple algebras and unirationality of conic bundles — 69

V. V. Kursov and V. I. Yanchevskiĭ, Crossed products of simple algebras and their automorphism groups — 75

A. S. Rapinchuk, On the metaplectic kernel for anisotropic groups — 81

A. S. Rapinchuk, The metaplectic kernel for the group $SL(1, D)$ — 87

A. S. Rapinchuk, On finite presentability of reduced norms in simple algebras — 93

A. Bovdi, Unitarity of the multiplicative group of a group algebra — 99

O. I. Tavgen', Profinite completions of linear solvable groups — 107

O. I. Tavgen', On the Grothendieck and Platonov conjectures — 113

A. A. Sharomet, The congruence problem for solvable algebraic groups over global fields of positive characteristic — 119

A. V. Prasolov, K-theory of free products and the Bass problem	125
A. V. Prasolov, On a theorem of Gersten for graded rings	131
A. V. Prasolov, Algebraic K-theory of Banach algebras	133
V. V. Benyash-Krivets, On the finite generation of the character ring of three-dimensional unimodular group representations	139
S. A. Baĭramov, The K_G-functor on the category of inverse spectra of topological spaces	145
M. R. Bunyatov and S. A. Baĭramov, K-theory on the category of distributive lattices	153
M. R. Bunyatov and S. A. Baĭramov, K-theory on the category of topological spaces	159
M. R. Bunyatov and S. A. Baĭramov, K-theory on the category of Boolean algebras with closure	165
M. R. Bunyatov and S. A. Baĭramov, On index theory for a family of Fredholm complexes	171
A. M. Bikchentaev, On noncommutative function spaces	179
A. V. Shtraus, Generalized resolvents of nondensely defined bounded symmetric operators	189

Russian Contents*

В. П. Платонов и А. А. Шаромет, О конгруэнц-проблеме для линейных групп над арифметическими кольцами, Докл. Акад. наук БССР, 1972, Том XVI, №. 5, 393–396

В. П. Платонов, Бирациональные свойства приведенной группы Уайтхеда, Докл. Акад. наук БССР, 1977, Том XXI, №. 3, 197–198

В. А. Артамонов, Проективные модули над групповыми кольцами нильпотентных групп, «Алгебра». Сб. работ, посвящ. 90-летию со дня рождения О. Ю. Шмидта. Москва, 1982, 7–23

С. Б. Востоков и И. Б. Фесенко, О кручении в высших функторах Милнора многомерных локальных полей, «Кольца и модули. Предел. теоремы теории вероятностей. Вып. 1», Ленинград, 1986, 75–87

А. А. Клячко, K-теория моделей Демазюра, «Исслед. по теории чисел» (Саратов), 1982, №. 8, 61–72

В. А. Липницкий, Нормы в полях вещественных алгебраических функций и приведенная группа Уайтхеда, Докл. Акад. наук БССР, 1982, Том XXVI, № 7, 585–588

В. А. Липницкий, Приведенная группа Уайтхеда слабо и вполне разветвленных тел, Весци Акад. навук БССР Сер. Физ. Мат. Навук, 1986, № 2, 28–30

В. И. Каскевич, Строение унитарных и ортогональных групп над гензелевыми дискретно нормированными алгебрами Адзумая, Докл. Акад. наук БССР, 1987, Том XXXI, № 12, 1073–1076

---

*The American Mathematical Society scheme for transliteration of Cyrillic may be found at the end of index issues of *Mathematical Reviews*.

В. И. Знчевский, Приведенные унитарные группы Уайтхеда и некоммутативные рациональные функции, Докл. Акад. наук БССР, 1980, Том XXIV, № 7, 588–591

И. И. Воронович и В. И. Янчевский, Рациональные поля разложения простых алгебр и унирациональность расслоений на коники, Докл. Акад. наук БССР, 1986, Том XXX, № 4, 293–296

В. В. Курсов и В. И. Янчевский, Скрещенные произведения простых алгебр и их групп автоморфизмов, Докл. Акад. наук БССР, 1988, Том XXXII, № 9, 777–780

А. С. Рапинчук, О метаплектическом ядре для анизотропных групп, Докл. Акад. наук БССР, 1985, Том XXIX, № 12, 1068–1071

А. С. Рапинчук, Метаплектическое ядро для группы $SL(1,D)$, Докл. Акад. наук БССР, 1986, Том XXX, № 3, 197–200

А. С. Рапинчук, О конечной определяемости приведенных норм в простых алгебрах, Докл. Акад. наук БССР, 1988, Том XXXII, № 1, 5–8

А. Бовди, Унитарность мультипликативной группы групповой алгебры, Уч. зап. Тартуск. ун-та, 1987, 764, 3–11

О. И. Тавгень, Проконечные пополнения линейных разрешимых групп, Докл. Акад. наук БССР, 1987, Том XXXI, № 6, 485–488

О. И. Тавгень, О гипотезах Гротендика и Платонова, Докл. Акад. наук БССР, 1988, Том XXXII, № 6, 489–492

А. А. Шаромет, Конгруэнц-проблема для разрешимых алгебраических групп над глобальными полями положительной характеристики, Докл. Акад. наук БССР, 1987, Том XXXI, № 3, 201–204

А. В. Прасолов, K-теория свободных произведений и проблема Басса, Докл. Акад. наук БССР, 1981, Том XXV, № 7, 598–600

А. В. Прасолов, О теореме Герстена для градуированных колец, Вестник БГУ им. В. И. Ленина, 1981, сер. 1, № 2, 62–63

А. В. Прасолов, Алгебраическая K-теория банаховых алгебр, Докл. Акад. наук БССР, 1984, Том XXVIII, № 8, 677–679

В. В. Беняш-Кривец, О конечной порожденности кольца характеров трехмерных унимодулярных представлений групп, Докл. Акад. наук БССР, 1986, Том XXX, № 5, 397–399

С. А. Байрамов, K_G-функтор на категории обратных спектров топологических пространств, Известия Акад. наук Азербайджанской ССР Сер. Физ.-Техн. Мат. наук, 1979, № 2, 3–9

М. Р. Бунятов и С. А. Байрамов, K-теория на категории дистрибутивных решеток, Докл. Акад. наук Азербайджанской ССР, 1983, Том XXXIX, № 5, 7–11

М. Р. Бунятов и С. А. Байрамов, K-теория на категории топологических пространств, Докл. Акад. наук Азербайджанской ССР, 1983, Том XXXIX, № 6, 14–18

М. Р. Бунятов и С. А. Байрамов, K-теория на категории Булевых алгебр с замыканием, Докл. Акад. наук Азербайджанской ССР, 1977, Том XXXIII, № 12, 3–7

И. П. Бунятов и С. А. Байрамов, K теории индекса семейства фредгольмовых комплексов, «Вопр. геометрии и алгебр. топол.» Баку, 1985, 56–76

А. М. Бикчентаев, О некоммутативных функциональных пространствах, «Функциональный анализ. Спектральная теория», Межвуз. сб. научн. тр., Ульяновск, 1987, 33–43

А. В. Штраус, Обобщенные резольвенты неплотно заданных ограниченных симметрических операторов, «Функциональный анализ. Спектральная теория», Межвуз. сб. научн. тр., Ульяновск, 1987, 187–196

On the Congruence Problem for Linear Groups over Arithmetic Rings

UDC 513.6

V. P. PLATONOV AND A. A. SHAROMET

In [1] one of the authors of the present article proved that the congruence problem always has a positive solution for solvable linear groups over integers.

The goal of the present article is to generalize the results of [1] to the case of solvable groups over arithmetic rings. Namely, let k be a field of algebraic numbers, and V the set of all non-Archimedean (nonequivalent) valuations of k; for an arbitrary finite subset $S \subset V$ let $O(S)$ denote the subring of k consisting of the elements that are integral for all $v \in V\setminus S$. The rings $O(S)$ are commonly called arithmetic rings.

Let $\Gamma \subset \mathrm{GL}(n, Z(S))$ be a linear group over $O(S)$. Recall that a subgroup $H \subset \Gamma$ is called a congruence subgroup of Γ if $H \supset \Gamma \cap \mathrm{GL}(n, \mathfrak{p}, O(S))$, where $\mathrm{GL}(n, \mathfrak{p}, O(S))$ is the principal congruence subgroup of $\mathrm{GL}(n, O(S))$ corresponding to the ideal $\mathfrak{p} \in O(S)$.

We refer to [1] and its bibliography for all necessary information about the congruence problem. We remark only that the standard construction of the ground field restriction enables us to reduce the investigation of the congruence problem to the case when $O(S)$ is an arithmetic subring of the field Q of rational numbers. Then S can be regarded as a finite set $\{p_1, \dots, p_k\}$ of prime numbers, and $O(S) = Z(S) \subset Q$ represents the set of all rational numbers whose denominators have prime divisors only in S.

For an arbitrary group G let F_G denote some subgroup of finite index.

Let $\Gamma \subset \mathrm{GL}(n, Z(S))$ be a solvable subgroup. Denote by Γ_U its unipotent part. As usual, $c(\Gamma)$ is the congruence kernel of the group Γ; $\dim \Gamma_U$ is the dimension of Γ_U in the Zariski topology.

MAIN THEOREM. *The congruence problem has a positive solution for Γ (i.e., $c(\Gamma) = (e)$) if and only if Γ_U is an S-arithmetic group. In the general*

1991 *Mathematics Subject Classification.* Primary 19B37, 20G30.
Translation of Dokl. Akad. Nauk BSSR **16** (1972), no. 5, 393–396.

case a subgroup $H \subset \Gamma$ of finite index is a congruence subgroup if and only if $(m, p_i) = 1$ $\forall p_i \in S$, where $m = [\Gamma_U : H \cap \Gamma_U]$. The congruence kernel satisfies $c(\Gamma) = \prod_{p_i \in S} N_{p_i}$, where N_{p_i} is a Sylow pro-p_i-subgroup of the group $\hat{\Gamma}$ and is a p_i-adic nilpotent Lie group with $\dim N_{p_i} \leq \dim \Gamma_U$.

The proof of the main theorem is in essence analogous to that of Theorem 1 in [1] and is based on the following general assertion, which strengthens Theorem 2 in [1].

THEOREM 1. *Let G be an algebraic Q-group, and F an arbitrary maximal semisimple Q-subgroup of G. Then $c(G_{Z(S)}) = c(F_{Z(S)})$; in particular, if G is a solvable group, then $c(G_{Z(S)}) = (e)$.*

PROOF. Suppose first that the group G is a Q-semidirect product of a normal subgroup H and a subgroup D. Then it is proved in a way completely analogous to that in [1] that for every m with $(m, p_i) = 1$ $(p_i \in S)$ the product $D_{Z(S)}(m) H_{Z(S)}(m)$ of S-congruence subgroups modulo m is an S-congruence subgroup of $G_{Z(S)}$. If $g \in c(G_{Z(S)})$, then $g = (g_1, \ldots, g_i, \ldots)$, where $g_i \in G_{Z(S)}$ and $\lim_{i \to \infty} g_i = e$ in the S-congruence topology of the group $G_{Z(S)}$. Then it can be assumed that $g_m \in D_{Z(S)}(m) H_{Z(S)}(m)$ and $g_m = d_m h_m$, where $d_m \in D_{Z(S)}(m)$ and $h_m \in H_{Z(S)}(m)$. This implies that $\lim_{i \to \infty} d_i = e$ and $\lim_{i \to \infty} h_i = e$ in the S-congruence topology, i.e., $h = (h_1, h_2, \ldots, h_i, \ldots) \in c(H_{Z(S)})$, $(d_1, d_2, \ldots, d_i, \ldots) \in c(D_{Z(S)})$, and $g = dh$. Consequently, $c(G_{Z(S)}) = c(D_{Z(S)}) c(H_{Z(S)})$.

Suppose now that G is an almost direct product over Q: $G = DH$, where H is an algebraic torus. Then $c(G_{Z(S)}) = c(D_{Z(S)})$. Indeed ([1], Lemma 2), it was actually proved that, as above, $D_{Z(S)}(m) H_{Z(S)}(m)$ is an S-congruence subgroup for any m with $(m, p_i) = 1$ $(p_i \in S)$. If $g = (g_1, g_2, \ldots, g_i, \ldots) \in c(G_{Z(S)})$, then, as above, $g = dh$, where $d \in c(D_{Z(S)})$ and $h \in c(H_{Z(S)})$. But $c(H_{Z(S)}) = (e)$ by Chevalley's theorem, hence $g = d \in c(D_{Z(S)})$, i.e., $c(G_{Z(S)}) = c(D_{Z(S)})$.

To complete the proof of Theorem 1 it is necessary to use the decomposition $G = FTU$, where TU is the solvable radical and U the unipotent radical of G, and to use the assertions proved above after employing the fact that $c(U_{Z(S)}) = (e)$ and $c(T_{Z(S)}) = (e)$.

We mention that in general $c(G_{Z(S)}) \neq c(D_{Z(S)}) c(H_{Z(S)})$ for the almost direct product $G = DH$.

The following obvious remark will be needed.

REMARK. Let G be an arbitrary group, H a subgroup of it, F_H a subgroup of finite index in H, and N a normal subgroup of G such that $N \cap H \subset F_H$. Then $F_G \cap HN \subset F_H N \Rightarrow F_G \cap H \subset F_H$.

LEMMA 1. *Suppose that H is an arbitrary subgroup of the group $A = (Z(S)^+)^n$ that is the direct sum of n copies of the additive group $Z(S)^+$. If* for $m = [H: F_H]$ the condition $(m, p_i) = 1$ holds for all $p_i \in S$, then there is an F_A such that $F_A \cap H \subset F_H$.

PROOF. It can be assumed that $H \neq (0)$.

a) Suppose that $n = 1$, i.e., $A = Z(S)^+$. Then $[Z^+ : H \cap Z^+] < \infty$, since $H \cap Z^+ \neq (0)$, hence $[HZ^+ : H] < \infty$ and $\exists a = \prod_{p_i \in S} p_1^{\alpha_i}$ ($\alpha_i > 0$ integers) such that $(t, p_i) = 1$ $\forall p_i \in S$ for $t = [H(aZ^+) : H]$. Consequently, it can be assumed that $H \supset aZ^+$. Then A/H is a periodic S-group. Let $A(m)$ be the principal congruence subgroup of A modulo m, and let $H(m) = H \cap A(m)$. Then $F_H \supset mH = H(m)$. The last equality follows from the fact that A/H does not contain elements of order m.

b) $n > 1$. $[H \cap A_i : F_H \cap A_i] < \infty$, where $A_i = Z(S)^+$ is the ith direct summand for A; then by using the remark and part a) it is possible to reduce the proof to the obvious case $[A : H] < \infty$.

LEMMA 2. *Suppose that H is an arbitrary subgroup of the group $G_{Z(S)}$, where G is a solvable algebraic Q-group, and F_H is a subgroup of finite index in H. Then the following assertions are equivalent*:

i) *there is an $F_{G_{Z(S)}}$ such that $H \cap F_{G_{Z(S)}} \subset F_H$;*
ii) $(m, p_i) = 1$ *for all $p_i \in S$, where $m = [H_U : H_U \cap F_H]$.*

PROOF. Let $U = G_U$.

1. i) $\Rightarrow$ ii). Since $c(G_{Z(S)}) = (e)$ by Theorem 1, there exists an ideal $I \subset Z(S)$ such that $H \cap U(I) \subset H_U \cap F_H$. Therefore, it suffices to observe that $d = [U_{Z(S)} : U(I)]$ is not divisible by any $p_i \in S$. Indeed, suppose that $I \neq a = [Z(S) : I]$; obviously, $(a, p_i) = 1$ $\forall p_i \in S$. Let $u \in U_{Z(S)}$, $u = 1 + x$, where $x^n = 0$. Then $u^{a^n} \in U(I)$, and hence $(d, p_i) = 1$ $\forall p_i \in S$.

2. ii) $\Rightarrow$ i). The lemma will be proved if we show that there exists an integer $t \neq 0$ such that $U(t) \cap H \subset F_H$. Indeed, then in view of the remark we can assume that $F_H \supset U(t)$ and find a subgroup F of finite index in $G_{Z(S)}/U(t)$ such that $F \cap H/U(t) \subset F_H/U(t)$, but, as is not hard to see, $G_{Z(S)}/U(t)$ is a polycyclic group, and the lemma follows from Lemma 4 in [1]. Thus, it suffices to prove the lemma for the case $G = U$.

The proof will be by induction on the class of nilpotency.

a) U Abelian. Then it is Q-isomorphic to a finite direct sum of additive groups of fields. Hence, there exists an integer $a \neq 0$ such that $U(a)$ is isomorphic to a subgroup of $(Z(S)^+)^n$, and the lemma follows from Lemma 1.

b) Denote by $\varphi : U \to U/U'$ the natural Q-homomorphism of U onto the unipotent Abelian group U/U'. Then there exists an integer $a_1 \neq 0$ such that $\varphi(U(a_1)) \subset \varphi(U)_{Z(S)}$. Since the class of nilpotency of U' is smaller by 1 than that of U, there exists by the induction hypothesis an integer $a_2 \neq 0$ such that $U'(a_2) \cap H \subset F_H$. Hence, in view of the remark it can be assumed that $F_H \supset U'(a_2)$. If $a = a_1 a_2$, then $\varphi(U(a)) \subset \varphi(U)_{Z(S)}$, and

$F_H \supset U'(a)$, and it suffices to find a subgroup F of finite index in $\varphi(U(a))$ such that $F \cap \varphi(H(a)) \subset \varphi(F(a))$. Such a subgroup exists in view of a), and the lemma is proved.

LEMMA 3. *Let A and H be the same as in Lemma 1. If for all $p \in S$ the group H does not have subgroups of index a multiple of p, then H is S-arithmetic.*

PROOF. We use induction on the number of direct summands $A_i = Z(S)^+$.

a) $A = Z(S)^+$. Then H is S-arithmetic if and only if $[A:H] < \infty$, but if $[A:H] = \infty$, then there exists an F_H such that $[H:F_H] = p$ for some $p \in S$.

b) Suppose that $\pi_i: A \to A_i$ is the natural projection of A onto A_i, and $\overline{\pi}_i$ is the projection of A onto $A_1 \oplus \cdots \oplus A_{i-1} \oplus A_{i+1} \oplus \cdots \oplus A_n$. If there exists an i such that $\operatorname{Ker} \pi_i = 0$ or $\operatorname{Ker} \overline{\pi}_i = 0$, then we get that H is isomorphically imbedded over Q in $(Z(S)^+)^m$ with $m < n$. If $\operatorname{Ker} \overline{\pi}_i \neq 0$, for all i, then they must be S-complete subgroups of A_i, and hence $[A_i: \operatorname{Ker} \overline{\pi}_i] < \infty$. Consequently, $[A:H] < \infty$, since $H \supset \operatorname{Ker} \overline{\pi}_1 \oplus \operatorname{Ker} \overline{\pi}_2 \oplus \cdots \oplus \operatorname{Ker} \overline{\pi}_n$. The lemma is proved.

Suppose that $U \subset \operatorname{GL}(n, Z(S))$ is a unipotent group; then for it we have

LEMMA 4. *U is S-arithmetic if and only if it does not contain subgroups with $p \in S$ in their indices.*

PROOF. The necessity was actually obtained in the proof of Lemma 2.

The sufficiency will be proved by induction on the nilpotency class of U.

a) U Abelian. Then there exists a congruence subgroup that is Q-isomorphic to a subgroup of $(Z(S)^+)^n$. It is S-arithmetic by Lemma 3.

b) U' satisfies the condition of the lemma, hence U' is an S-arithmetic group by the induction hypothesis. Then there exists an integer $a_1 \neq 0$ such that $U' \supset \overline{U}'(a_1)$, where $\overline{}$ is the closure in the Zariski topology. Let $\varphi: \overline{U} \to \overline{U}/\overline{U}'$ be the natural Q-homomorphism of U onto the corresponding Abelian unipotent group. Then $\varphi(U(a_1))$ is S-arithmetic in view of a), hence there exists an $a_2 \neq 0$ such that $\varphi(U(a_1)) \supset \varphi(\overline{U})(a_2)$. It is known that there exists an $a = a_1 t$ such that $\varphi(\overline{U}(a)) \subset \varphi(\overline{U})(a_2)$. It is now not hard to see that $\overline{U}(a) \subset U$, i.e., U is an S-arithmetic group.

Lemmas 2 and 4 amount to an exhaustive characterization of subgroups of finite index that are congruence subgroups. For completeness we compute also the congruence kernel.

Let $\Gamma \subset \operatorname{GL}(n, Z(S))$ be a solvable subgroup. If F_Γ is a normal subgroup of Γ, then there is a subgroup $A \subset \Gamma_U$ such that $F_\Gamma A$ is a congruence subgroup. Moreover, by analogy with the proof of Lemma 2 we get that $\widehat{\Gamma} \supset \widehat{\Gamma}_U$. This implies that $c(\Gamma) = c(\Gamma_U)$.

The groups $\widehat{\Gamma}_U$ and $\overline{\Gamma}_U$ are pronilpotent, hence they can be decomposed into direct products of their Sylow pro-p-groups:

$$\widehat{\Gamma}_U = \prod_{p \in S} N_p \times \prod_{p \notin S} N_p, \qquad \overline{\Gamma}_U = \prod_{p \notin S} N_p. \qquad (1)$$

The second equality follows from the fact that the congruence topology on Γ_U coincides with the topology of subgroups of finite index that is not divisible by any $p \in S$. It follows from (1) that $c(\Gamma_U) = \prod_{p \in S} N_p$. In conclusion we present a lemma that finishes the proof of the main theorem.

LEMMA 5. *If $U \subset \mathrm{GL}(n, Z(S))$ is a unipotent group, then its p-completion N_p is a p-adic nilpotent Lie group, and $\dim N_p \leq \dim U$.*

SKETCH OF PROOF. Let $M = \bigcap_{\alpha=1}^{\infty} U^{p^\alpha}$, where U^m is the subgroup generated by the mth powers. Then $M = U \cap \overline{M}$. Hence, $\varphi \colon U \to U/M$ is a Q-rational homomorphism, and $\varphi(U)$ is a unipotent group such that $\bigcap_{\alpha=1}^{\infty} \varphi(U)^{p^\alpha} = (e)$. It is not hard to see that N_p coincides with the closure of $\varphi(U)$ in the p-adic topology.

The proof of the main theorem follows from Theorem 1 and Lemmas 2, 4, and 5.

BIBLIOGRAPHY

1. V. P. Platonov, *The congruence problem for solvable integral groups*, Dokl. Akad. Nauk BSSR **15** (1971), no. 6, 869–872. (Russian)

Institute of Mathematics
Academy of Sciences of the Belorussian SSR

Received 28/DEC/71

Translated by H. H. McFADEN

PROOF. Denote by d the dimension of the variety G. By assumption, there exist a K-defined d-dimensional affine space V and a birational K-isomorphism $\varphi: V \to G$. In other words, the field $K(G)$ of K-defined rational functions on G is isomorphic to the field $K(x_1, \ldots, x_d) = F$. Let us consider $A_F = A \otimes_K F$. If ω is a generic point of G over K, then $K(\omega) \cong F$. Therefore, it can be assumed that $\omega \in SL(1, A_F)$. Since $SK_1(A)$ is a group of finite exponent for any A (see [2], Lemma 2.1), the generic point satisfies $\omega^r \in [A_F^*, A_F^*]$ for some r. Thus, we can choose a generic point ω in $[A_F^*, A_F^*]$. By the stability theorem in [3], $SK_1(A) \cong SK_1(A_F) \cong SK_1(A_{K(\omega)})$, therefore, it can be assumed that $F = K(\omega)$. Let $\omega = [a_1, a_2] \cdot [a_3, a_4] \cdots [a_{2t-1}, a_{2t}]$ be an expression for ω in terms of the commutators, with smallest possible t. If $\omega = (\omega_i)$ in the corresponding coordinate system, then $\omega_i = \varphi_i(x_1, x_2, \ldots, x_d) \in F$. Since $\varphi_K: V_K \to G_K$ is relatively open, all possible specializations $x_i \to \theta_i \in K$ form a relatively open subset θ of G_K. Every a_i is in A_F, and $a_i = (f_{ik})$, $a_i^{-1} = (g_{ik})$, $f_{ik}, g_{ik} \in F$. Denote by h the product of the denominators of all the rational functions $\varphi_i, f_{ik}, g_{ik}$. If U_h is a principal open subset of V_K corresponding to h, then the specializations $a_i(U_h)$ and $a_i^{-1}(U_h)$ of the respective points a_i and a_i^{-1} induce by means of a product of commutators a certain set R of specializations of the generic point ω. It is not hard to see that $R = \varphi_K(U_h) \cap \theta$, and R is relatively open, with every element of R representable by construction as a product of t commutators in A^*. Then $R^2 = G_K$, and every element in G_K is a product of at most $m = 2t$ commutators, in particular, $SK_1(A) = 1$. The theorem is proved.

The results on $SK_1(A)$ obtained by the author show that, as a rule, $SK_1(A) \neq 1$ over sufficiently general fields. Therefore, it follows from the theorem that the rationality of the variety of a simply connected algebraic group G corresponding to $G_K = SL(1, A)$ is a comparatively rare phenomenon. It is rather surprising that previous to the present note hardly anything was known about rationality of such natural varieties of $SL(1, A)$. However, the surprise passes if we take into account the discovered connection with triviality of the reduced Whitehead group—a problem having a considerably longer history (see [2]), and also the general difficulties associated with problems of Lüroth type (see [4]). At the same time, several extremely natural questions await their solutions in the near future. The most important and difficult among them is to determine the condition $SK_1(A) = 1$ implies the rationality([1]) of $SL(1, A)$. In other words, what are sufficient tests for the rationality of the norm varieties of $SL(1, A)$? The answer is not known even for the case of a field K of algebraic numbers. We mention separately

---

([1])*Added in print*. It follows from [2] that even for natural fields (for example, fields of rational functions) there exist nonrational varieties of $SL(1, A)$ with invariant $SK_1(A) = 1$, therefore, the class of fields K for which $SK_1(A) = 1$ implies that $SL(1, A)$ is rational cannot be broad.

also the case of algebras A of prime index: here $SK_1(A) = 1$, and it can be assumed that the variety of $SL(1, A)$ is rational.

BIBLIOGRAPHY

1. V. P. Platonov, *On the Tannaka-Artin problem*, Dokl. Akad. Nauk SSSR **221** (1975), 1038–1041; English transl. in Soviet Math. Dokl. **16** (1975).
2. ____, *The Tannaka-Artin problem, and reduced K-theory*, Izv. Akad. Nauk SSSR Ser. Mat. **40** (1976), no. 2, 227–261; English transl. in Math. USSR-Izv. **10** (1976).
3. ____, *Reduced K-theory and approximation in algebraic groups*, Trudy Mat. Inst. Steklov. **142** (1976), 198–207; English transl. in Proc. Steklov Inst. Math. **142** (1979).
4. V. E. Voskresenskiĭ, *Fields of invariants for Abelian groups*, Uspekhi Mat. Nauk **28** (1973), no. 4, 77–102; English transl. in Russian Math. Surveys **28** (1973).

Institute of Mathematics
Academy of Sciences of the Belorussian SSR

Received 5/NOV/76

Translated by H. H. McFADEN

Projective Modules over Group Rings of Nilpotent Groups

V. A. ARTAMONOV

In this paper we describe the finitely generated infinite nilpotent groups that have integral group rings over which all projective modules are free. The principal results are Theorems 4–6. In particular, it is proved that these nilpotent groups are abelian. We note in this regard the description of the finite groups that have integral group rings over which all projective modules are free. For these groups G the class group $\text{Cl}(\mathbb{Z}G)$ is trivial; moreover, for an abelian group G, triviality of the class group implies that all projective modules over $\mathbb{Z}G$ are free. It was shown in [1] that for a finite abelian group G, the class group $\text{Cl}(\mathbb{Z}G)$ is trivial if and only if G is either7 a group of order 4 or a cyclic group of order 1–11, 13, 14, 17, 19. If G is a nonabelian finite group with $\text{Cl}(\mathbb{Z}G)$ trivial, then G is one of the following groups [2]: 1) the dihedral group $\mathscr{D}_n$, $n \geq 3$; or 2) A_4, S_4, A_5. Conversely ([3], pp. 25, 31), the class groups $\text{Cl}(\mathbb{Z}\mathscr{D}_3)$, $\text{Cl}(\mathbb{Z}\mathscr{D}_4)$, $\text{Cl}(\mathbb{Z}A_4)$, $\text{Cl}(\mathbb{Z}S_4)$, and $\text{Cl}(\mathbb{Z}A_5)$ are trivial.

In the study of projective modules over group rings of infinite groups, free abelian groups were considered first. A. A. Suslin [4] and R. G. Swan [5] showed that if k is a principal ideal domain and G is a free abelian group of finite rank, then all projective kG-modules are free. The situation becomes different for infinite nonabelian groups. In [6] and [7] examples were constructed of groups G such that there exist a countable number of nonisomorphic left ideals P in $\mathbb{Z}G$ with $P \oplus \mathbb{Z}G \simeq \mathbb{Z}G \oplus \mathbb{Z}G$. It was shown in [8] that if the nonabelian group G is the union of a countable subnormal series

$$1 = G_0 \subseteq G_1 \subseteq \cdots \subseteq G_n \subseteq \cdots$$

with free abelian factors, and k is a commutative ring, then there exists a

---

1980 *Mathematics Subject Classification* (1985 Revision). Primary 13C10, 20C07; Secondary 16A27, 19A31, 20F18.

Translation of Algebra, Moskov. Gos. Univ., Moscow, 1982, pp. 7–23.

nonfree left ideal P in kG such that

$$P \oplus kG \simeq kG \oplus kG.$$

If k is a Noetherian domain in which each prime number is noninvertible, then there exist a countable number of left ideals P in kG having these properties that are nonisomorphic as modules. In particular, all of these results are true (by [9], Theorem 17.2.2) for finitely generated torsion-free nilpotent groups. We note that by [10], for a polycyclic group G that is the union of a subnormal series of the form mentioned above, $K_0(\mathbb{Z}G) \simeq \mathbb{Z}$ (cf. also [11]).

Several important results were obtained in the paper [12]. It was shown, for example, that if k is a field of positive characteristic or $k = \mathbb{Z}$, G is a polycyclic-by-finite group, and h is the polycyclic rank of G, then for each finitely generated kG-module M we have $M \simeq (kG)^r \oplus N$, where the quasirank of N is not more than $H + \operatorname{Krull dim} k$. The Cancellation Theorem is valid. If R is the Artinian full ring of quotients of the ring kG, and P is a finitely generated projective kG-module, then the R-module $R \otimes_{kG} P$ is free and its rank coincides with the quasirank of P. Let P be a finitely generated projective kG-module, and let k be as above or a field of finite transcendence degree over the field of rational numbers. Then there exists a normal subgroup N of finite index in G such that P is a free kN-module.

We also mention that the following result was announced in [13]. If G is a polycyclic group with two generators, and one relation not in the commutator subgroup, then there exists a nonfree projective $\mathbb{Q}G$-module.

All rings considered in the paper are associative and have an identity. If x is an invertible element of the ring R and $y \in R$, then $y^x = x^{-1}yx$. If M is a subset of R, then

$$M^x = \{m^x | m \in M\}.$$

All definitions involving group rings can be found in [14].

§1

This section is of a preparatory nature. We keep the following notation fixed throughout the whole section: G is a group, H is a normal subgroup of G, and k is a commutative ring.

PROPOSITION 1. *Let M be a system of representatives of the cosets of H in G. Then each element $z \in kG$ is uniquely representable in the form*

$$z = \sum g z_g, \tag{1}$$

where $z_g \in kH$, and g runs through M.

PROOF. Each element $x \in G$ is uniquely representable in the form $x = gh$, where $g \in M$ and $h \in H$. Consequently, each element from kG has a

representation of the form (1). Now let
$$\sum_{g \in M} g z_g = 0,$$
$$z_g = \sum_{h \in H} z_{g,h} h \in kH, \qquad z_{g,h} \in k.$$
Then
$$\sum_{g \in M, h \in H} z_{g,h} gh = 0. \tag{1'}$$
If $gh = g'h'$, then $g = g' \in M$ and $h = h' \in H$. Therefore in (1') we obtain $z_{g,h} = 0$, so that $z_g = 0$ for all $g \in M$.

PROPOSITION 2. *With M as in Proposition 1, if $\mathscr{L}$ is a left ideal in kH, then z in (1) lies in $kG\mathscr{L}$ if and only if $z_g \in \mathscr{L}$ for all $g \in M$. In particular, $kG\mathscr{L} \cap kH = \mathscr{L}$. If $\mathscr{L}^x = \mathscr{L}$ for all $x \in M$, then $kG\mathscr{L}$ is an ideal in kG.*

PROOF. Let $z \in kG$ have the representation (1) and $r \in \mathscr{L}$. Then zr has the representation
$$zr = \sum_{g \in M} g(z_g r), \qquad z_g r \in \mathscr{L}.$$
We obtain all of the assertions from this and the uniqueness of the representation (1). The proposition is proved.

We now assume that the group G/H is well ordered. This induces a well-ordering on M. If $\mathfrak{p}$ is an ideal in kH and $z \in kG \backslash kG\mathfrak{p}$ has the representation (1), then the monomial gz_g with greatest (least) $g \in M$ among those whose $z_g \in kH \backslash \mathfrak{p}$ will be called the $\mathfrak{p}$-highest ($\mathfrak{p}$-lowest) term of z. These monomials are well defined, by Proposition 2.

PROPOSITION 3. *Let $\mathfrak{p}$ be a prime ideal in kH, so that $kH/\mathfrak{p}$ has no zero divisors, with $\mathfrak{p}^x = \mathfrak{p}$ for all $x \in M$. Then the $\mathfrak{p}$-highest ($\mathfrak{p}$-lowest) term of a product of elements from $kG \backslash kG\mathfrak{p}$ is equal to the product of the $\mathfrak{p}$-highest ($\mathfrak{p}$-lowest) terms of the factors.*

PROOF. Let $t = \sum_{g \in M} g t_g$ and $z = \sum_{g \in M} g z_g$ have $\mathfrak{p}$-highest ($\mathfrak{p}$-lowest) terms yt_y and xz_x, respectively. Then
$$tz = \sum_{g,d \in M} g t_g d z_d = \sum_{g,d \in M} gd t_g^d z_d.$$
Since G/H is fully ordered and $kH/\mathfrak{p}$ has no zero divisors, the $\mathfrak{p}$-highest ($\mathfrak{p}$-lowest) term of tz is equal to $yt_y x z_x$, by Proposition 1.

PROPOSITION 4. *Let $\mathfrak{p}$ be a prime ideal in kH, with $\mathfrak{p}^x = \mathfrak{p}$ for all $x \in M$. Assume that $t \in kG$ is invertible in $kG/kG\mathfrak{p}$. Then $t \equiv gv \mod kG\mathfrak{p}$, where $g \in M$, $v \in kH$, and v is invertible in $kH/\mathfrak{p}$.*

PROOF. Let d be the inverse of t modulo $kG\mathfrak{p}$. Then $dt = 1 + u$, where $u \in kG\mathfrak{p}$. By Proposition 3, the $\mathfrak{p}$-highest term in t coincides with the $\mathfrak{p}$-lowest. Therefore $t \equiv gt_g \bmod kG\mathfrak{p}$, where $g \in M$ and $t_g \in kH$. Analogously, we also have $d \equiv g^{-1} d_{g^{-1}} \bmod kG\mathfrak{p}$, where $d_{g^{-1}} \in kH$. Hence $d_{g^{-1}}^g t_g = 1$ in $kH/\mathfrak{p}$. It remains to put $v = t_g$.

PROPOSITION 5. *Let q be an integer and let $\mathfrak{p}_i$, $i \in I$, be a system of ideals in kH with the following properties*: 1) *Each ring $kH/\mathfrak{p}_i$ has no zero divisors and $\mathfrak{p}_i^z = \mathfrak{p}_i$ for all $z \in M$.* 2) *If $\mathfrak{m}$ is the fundamental ideal in kH, then $kHq + \mathfrak{m} + \mathfrak{p}_i$ is a proper ideal in kH for all $i \in I$. Put $\mathfrak{q} = \bigcap_i \mathfrak{p}_i$ and assume that $f \in G \backslash H$, $z \in kG\mathfrak{m}$, and $h = 1 + qf + z$ is invertible in $kG/kG\mathfrak{q}$. Then $h \in kH + kG\mathfrak{q}$.*

PROOF. By Proposition 4, for each $i \in I$ we have $h = 1 + qf + z \in g_i v_i + kG\mathfrak{q}$, where $g_i \in M$, and v_i is invertible in $kH/\mathfrak{p}_i$. We assume that the system of representatives M contains 1 and f. Let $z = \sum t z_t$ be an expansion of type (1), where $t \in M$ and $z_t \in \mathfrak{m}$. Taking Propositions 1 and 2 into account, we find that one of the following cases is possible in $kH/\mathfrak{p}_i$: $z_t \equiv v_i$, $t \equiv g_i \bmod H$; $q + z_f \equiv v_i$, $f \equiv g_i \bmod H$; $1 + z_t \equiv v_i$, $t \equiv g_i \equiv 1 \bmod H$. But because of property 2), the first and second cases are impossible. Consequently, $g_i \in H$ and $h \in kH + kG\mathfrak{p}_i$. Let $h = \sum_{t \in M} th_t$. By what was proved above, we obtain $h_t \in \mathfrak{p}_i$ for all $i \in I$ when $t \neq 1$. Hence $h_t \in \mathfrak{q}$ and $h \in kH + kG\mathfrak{q}$.

§2

This section is also of a preparatory nature. We keep the following notation fixed throughout the whole section: $\mathscr{U}$ is a finitely generated infinite nilpotent group, whose torsion subgroup T contains $\mathscr{U}'$ and is a cyclic group $\{x\}$ of prime power order p^k. In this case $\mathscr{U}/T$ is a free abelian group. Let $t_1, \ldots, t_n$ be a complete set of free generators for the group $\mathscr{U}/T$. If $r \leq k$, $m = p^r$, and $\Phi_m(X)$ is the cyclotomic polynomial, then

$$\Phi_m(x) = 1 + x^{p^{r-1}} + \cdots + x^{(p-1)p^{r-1}}.$$

Therefore for all $g \in \mathscr{U}$ we have $\Phi_m(x)^g = \Phi_m(x^g) = \Phi_m(x)$, since $T = \{x\}$ is a normal subgroup of $\mathscr{U}$. Thus $\mathbb{Z}\mathscr{U}\Phi_m(x)$ is an ideal in $\mathbb{Z}\mathscr{U}$. Hence $\mathbb{Z}\mathscr{U}(1 + x + \cdots + x^{p^k - 1})$ is also an ideal in $\mathbb{Z}\mathscr{U}$.

The condition on the commutator subgroup $\mathscr{U}'$ makes it finite. Hence by [15, Corollary 2.6, p. 7], the center of $\mathscr{U}$ has finite index in $\mathscr{U}$. Thus there exists $s \geq 1$ such that $f = t_1^s$ is in the center of $\mathscr{U}$.

THEOREM 1. *For some $a \in \mathbb{Z}\mathscr{U}$ and integer d, let the element $h = 1 + dpf + a(x - 1)$ be invertible in the ring $\mathbb{Z}\mathscr{U}/\mathbb{Z}\mathscr{U}(1 + x + \cdots + x^{p^k - 1})$. Then $d = d'p^{k-1}$, $d' \in \mathbb{Z}$, and $h = 1 + fd'(1 + x + \cdots + x^{p^k - 1}) + a_0(x - 1)$, $a_0 \in \mathbb{Z}T$, where $1 + a_0(x - 1)$ is invertible in $\mathbb{Z}T$.*

PROOF. If $r \le k$ and $m = p^r$, then we put $\mathfrak{p}_m = \mathbb{Z}\mathscr{U}\Phi_m(x)$. We use Proposition 5 with $G = \mathscr{U}$, $H = T$, and $k = \mathbb{Z}$. We note that $\mathbb{Z}Tdp + \mathbb{Z}T(x-1) + \mathbb{Z}T\Phi_m(x) \subseteq \mathbb{Z}Tp + \mathbb{Z}T(x-1)$ is a proper ideal in $\mathbb{Z}T$, and $\mathfrak{q} = \bigcap \mathfrak{p}_m = \mathbb{Z}\mathscr{U}(1 + x + \cdots + x^{p^k-1})$. Thus by Proposition 5 we obtain $h = b + c(1 + x + \cdots + x^{p^k-1})$, where $c \in \mathbb{Z}\mathscr{U}$ and $b \in \mathbb{Z}T$ is invertible modulo $(1 + x + \cdots + x^{p^k-1})$. Let $c = \sum t_1^{i_1} \cdots t_n^{i_n} c_{i_1,\ldots,i_n}$ and $a = \sum t_1^{i_1} \cdots t_n^{i_n} a_{i_1,\ldots,i_n}$, where $i_1, \ldots, i_n \in \mathbb{Z}$ and $c_{i_1,\ldots,i_n}$, $a_{i_1,\ldots,i_n} \in \mathbb{Z}T$. Now equating in h the coefficients of $t_1^{i_1} \cdots t_n^{i_n}$ we obtain, by Proposition 1,

$$1 + a_{0,\ldots,0}(x-1) = b + c_{0,\ldots,0}(1 + x + \cdots + x^{p^k-1}),$$
$$dp + a_{s,0,\ldots,0}(x-1) = c_{s,0,\ldots,0}(1 + x + \cdots + x^{p^k-1}), \quad (2)$$
$$a_{i_1,\ldots,i_n}(x-1) = c_{i_1,\ldots,i_n}(1 + x + \cdots + x^{p^k-1}),$$

where in the last case $(i_1, \ldots, i_n) \ne (0, \ldots, 0), (s, 0, \ldots, 0)$. We note that in $\mathbb{Z}T$

$$(x-1) \cap (1 + x + \cdots + x^{p^k-1}) = 0.$$

Therefore it follows from (2) that $c = c_0 + fc_1$, $a = a_0 + fa_1$, where c_i, $a_i \in \mathbb{Z}T$, and $dp + a_1(x-1) = c_1(1 + x + \cdots + x^{p^k-1})$. But then in $\mathbb{Z}T$ we obtain

$$dp \in \mathbb{Z} \cap [(x-1) + (1 + x + \cdots + x^{p^k-1})] = (p^k).$$

Thus $dp = d'p^k$, where $d' \in \mathbb{Z}$. Hence

$$h = 1 = fd'(1 + x + \cdots + x^{p^k-1}) + (a_0 + fa_1')(x-1)$$
$$= b + (c_0 + fc_1)(1 + x + \cdots + x^{p^k-1}),$$

so that $a_1'(x-1) \in (1 + x + \cdots + x^{p^k-1}) \cap (x-1) = 0$.

Thus $h = 1 + fd'(1 + x + \cdots + x^{p^k-1}) + a_0(x-1)$, $a_0 \in \mathbb{Z}T$. It remains to show that $1 + a_0(x-1)$ is invertible in $\mathbb{Z}T$. We note that $1 + a_0(x-1)$ is invertible in $\mathbb{Z}T/(1 + x + \cdots + x^{p^k-1})$. Therefore there exists a $b_0 \in \mathbb{Z}T$ such that for some $v \in \mathbb{Z}T$

$$(1 + a_0(x-1))(1 + b_0(x-1))$$
$$= 1 + a_0(x-1) + b_0(x-1) + a_0 b_0(x-1)^2$$
$$= 1 + v(1 + x + \cdots + x^{p^k-1}).$$

But then

$$a_0(x-1) + b_0(x-1) + a_0 b_0(x-1)^2 \in (x-1) \cap (1 + x + \cdots + x^{p^k-1}) = 0.$$

Consequently, $(1 + a_0(x-1))(1 + b_0(x-1)) = 1$. The theorem is proved.

COROLLARY. *Under the conditions of Theorem 1 we have*

$$h = (1 + a_0(x-1))(1 + fd'(1 + x + \cdots + x^{p^k-1})),$$

where $1 + a_0(x-1)$ is invertible in $\mathbb{Z}T$.

The proof is obvious.

§3

This section, like the first two, is of a preparatory nature. Let $\mathscr{U}$ be a finitely generated infinite nilpotent group with torsion subgroup $T = \{x\} \times \{y\}$, a direct product of two cyclic groups of prime order p. In addition, let $\mathscr{U}' \subseteq T$. The intersection of the center of $\mathscr{U}$ with T is nontrivial. Therefore we may assume that y is in the center of $\mathscr{U}$. Moreover, as was noted in §2, the center has finite index in $\mathscr{U}$. Let the elements $t_1, \ldots, t_n$ freely generate the free abelian group $\mathscr{U}/T$, where $f = t_1^s$ is in the center of $\mathscr{U}$. We assume that $d \in \mathbb{Z}$, $a, v \in \mathbb{Z}\mathscr{U}$, and

$$h = 1 + dpf + a(x-1) + v(y-1)$$

is invertible in $\mathbb{Z}\mathscr{U}$ modulo $\mathbb{Z}\mathscr{U}\Phi_p(x)\Phi_p(y)$. Note that $\Phi_p(x)\Phi_p(y)$ is equal to the sum of all the elements in T. Factoring by the normal subgroup $\{y\}$, from the Corollary to Theorem 1 we obtain

$$h = (1 + a_0(x-1))(1 + d\Phi_p(x)f + v(y-1)),$$

where $1 + a_0(x-1)$ is invertible in $\mathbb{Z}\mathscr{U}$. Thus we find that

$$h' = 1 + d\Phi_p(x)f + v(y-1) \tag{3}$$

is invertible in $\mathbb{Z}\mathscr{U}$ modulo $\mathbb{Z}\mathscr{U}\Phi_p(x)\Phi_p(y)$.

THEOREM 2. *If h' in (3) is invertible modulo $\mathbb{Z}\mathscr{U}\Phi_p(x)\Phi_p(y)$, then p divides d.*

PROOF. We carry out the proof in several steps.

LEMMA 1. *Let t be an inverse of h' modulo $\mathbb{Z}\mathscr{U}\Phi_p(x)\Phi_p(y)$. Then $t = 1 - df\Phi_p(x) + w(y-1)$, $w \in \mathbb{Z}\mathscr{U}$, where $th' = h't = 1 - d^2f^2\Phi_p(x)\Phi_p(y)$.*

PROOF. We note that

$$(1 - dpf)(1 + dpf) = 1 - d^2p^2f^2.$$

Therefore by Theorem 1

$$t = 1 + b_0(x-1) - d\Phi_p(x)f + w(y-1),$$

where $b_0 \in \mathbb{Z}T$. In addition,

$$th' = 1 + c\Phi_p(x)\Phi_p(y).$$

Hence $th' \equiv 1 + b_0(x-1) - d^2f^2p\Phi_p(x) \equiv 1 + pc\Phi_p(x)$ modulo $(y-1)$, so that $b_0(x-1) = 0$ and $c = -d^2f^2$. The lemma is proved.

Thus,
$$th' = h't = [1 - df\Phi_p(x) + w(y-1)][1 + df\Phi_p(x) + v(y-1)] \\ = 1 - d^2f^2\Phi_p(x)\Phi_p(y). \tag{4}$$

We put
$$A(y-1) = pw(y-1) - df\Phi_p(x)(p - \Phi_p(y)), \\ B(y-1) = pv(y-1) + df\Phi_p(x)(p - \Phi_p(y)). \tag{5}$$

Multiplying (4) by p^2 we obtain
$$[p - df\Phi_p(x)\Phi_p(y) + A(y-1)][p + df\Phi_p(x)\Phi_p(y) + B(y-1)] \\ = p^2 - p^2d^2f^2\Phi_p(x)\Phi_p(y) + [p + A(y-1)][p + B(y-1)] - p^2 \\ = p^2 - p^2d^2f^2\Phi_p(x)\Phi_p(y).$$

Thus,
$$[p + A(y-1)][p + B(y-1)] = p^2. \tag{6}$$

Multiplying (6) on the left by $\Phi_p(xy^i)$ and on the right by $\Phi_p(xy^j)$ we obtain
$$[p\Phi_p(xy^i) + \Phi_p(xy^i)A(y-1)][p\Phi_p(xy^j) + B\Phi_p(xy^j)(y-1)] \\ = p^2\Phi_p(xy^i)\Phi_p(xy^j). \tag{7}$$

We choose as a system of representatives of the cosets of T in $\mathscr{U}$ elements $t_1^{r_1} \cdots t_n^{r_n}$, $r_i \in \mathbb{Z}$, where $t_1, \ldots, t_n$ is the basis of the free abelian group $\mathscr{U}/T$ selected earlier. Let $A = \sum gA_g$ and $B = \sum gB_g$ be expansions of type (1). Let us consider the product of the 0-highest (0-lowest) terms on the left side of (7):
$$[\Phi_p(xy^i)gA_g(y-1)][h\Phi_p(xy^j)B_h(y-1)] \\ = gh[\Phi_p(xy^i)^{gh}A_g^h(y-1)][B_h\Phi_p(xy^j)(y-1)]. \tag{8}$$

If $gh \neq 1$ in $\mathscr{U}/T$, then this product is equal to zero.

LEMMA 2. *The ring* $\mathbb{Z}T\Phi_p(xy^j)(y-1)$ *has no zero divisors.*

PROOF. It is sufficient to show that the ring $\mathbb{Q}T\Phi_p(x)(y-1)$ has no zero divisors. One can easily see that
$$\mathbb{Q}T\Phi_p(x)(y-1) = \mathbb{Q}T\left[\frac{1}{p}\Phi_p(x)\right]\left[\frac{1}{p}(p - \Phi_p(y))\right] \simeq \mathbb{Q}[\zeta],$$
where $\zeta = e^{2\pi i/p}$, from which the result follows.

Thus, taking into account that in (8) i and j are arbitrary and x is in the center of $\mathscr{U}/\{y\}$, we get from Lemma 2
$$A_g^h\Phi_p(xy^i)(y-1) = B_h\Phi_p(xy^j)(y-1) = 0$$

for all i, j. We note that $\operatorname{Ann}\Phi_p(xy^j) = (xy^j - 1)$ ([14], Chapter 1). Therefore

$$A_g^h(y-1), B_h(y-1) \in \bigcap_{j=0}^{p-1}(xy^j - 1). \qquad (8')$$

Hence

$$A_g(y-1), B_h(y-1) \in \left[\bigcap_{j=0}^{p-1}(xy^j - 1)\right](y-1), \qquad (8'')$$

since $[x, \mathscr{U}] \subseteq \{y\}$.

LEMMA 3. $(y-1)[\bigcap_{j=0}^{p-1}(xy^j - 1)] = 0$.

PROOF. It is sufficient to verify this in $\mathbb{Q}T$. If $\zeta = e^{2\pi i/p}$, then

$$\mathbb{Q}T = \mathbb{Q}e_0 \oplus F_1 \oplus \cdots \oplus F_t,$$
$$F_j \simeq \mathbb{Q}[\zeta], \qquad e_0 = \frac{1}{p^2}\Phi_p(x)\Phi_p(y). \qquad (8''')$$

We consider a homomorphism of $\mathbb{Q}T$ into some F_j or into $\mathbb{Q}$. By [16], Chapter III, §1.2, the image of T coincides with $\{\zeta\}$ or with 1. Consequently, the kernel of the homomorphism of T into $\{\zeta\}$ or into 1 contains either $y - 1$, or $xy^r - 1$ for some r. Hence we get the required assertion from $(8''')$.

From $(8'')$ and Lemma 3 we obtain $A_g(y-1) = B_h(y-1) = 0$. This proves:

LEMMA 4. *In* (6) *the 0-highest terms of* $A(y-1)$ *and* $B(y-1)$ *coincide with the 0-lowest terms and belong to* $\mathbb{Z}T$.

In particular, equating the coefficients of f, we get from equalities (5) that $d\Phi_p(x)(p - \Phi_p(y)) \equiv 0 \bmod p(y-1)$, whence p divides d. The theorem is proved.

§4

In this section we shall consider examples of projective modules over group rings.

THEOREM 3. *Let the group* G *contain a finite normal subgroup* H *of order* d. *Assume that* k *is a commutative ring and set*

$$N = \sum_{x \in H} x \in kG.$$

Let $f, g \in kG$, *with* $fg = gf = 1 + hd$, *where* $h \in kG$ *and* f, g *are not zero divisors in* kG/kGN. *Assume also that* d *is not a zero divisor in* k. *Then the left ideal* $P = kGf + kGN$ *is a projective left* kG-*module. If*

GROUP RINGS OF NILPOTENT GROUPS

the image of f in $(k/d)(G/H)$ is not the image of an invertible element in kG/kGN, then the module P is not free.

PROOF. We note that N is contained in the center of kG, since H is a normal subgroup of G. We put $S = kGg + kGN$ and consider the homomorphism of left kG-modules

$$\varphi: kGe_1 \oplus kGe_2 \to S \oplus P,$$
$$\varphi(e_1) = (g, hN), \quad \varphi(e_2) = (fg - hN, f). \tag{9}$$

Note that $xN = N$ for all $x \in H$. Therefore $N^2 = dN$. Hence

$$\varphi(fNe_1 - hNe_2) = (N, 0), \quad \varphi[(1 + hN)e_1 - hgNe_2] = (g, 0).$$

Thus from (9) we get $S \subseteq \operatorname{Im}\varphi$ and $(0, hN), (0, f) \in \operatorname{Im}\varphi$. But then $P \subseteq \operatorname{Im}\varphi$, since $N = (gN)f - d(hN)$. Thus φ is an epimorphism of left kG-modules.

Let us find $\operatorname{Ker}\varphi$. Let $a, b \in kG$ and $z = ae_1 + be_2 \in \operatorname{Ker}\varphi$. Then

$$ag + b(fg - hN) = 0, \quad ahN + bf = 0. \tag{10}$$

We note that $fh = hf$ and $gh = hg$, since $gf = fg = 1 + hd$ and d is not a zero divisor in kG. In addition

$$\begin{vmatrix} g & fg - hN \\ hN & f \end{vmatrix} = fg - fghN + (hN)^2 = 1 + h(d - N).$$

Hence from (10)

$$a[1 + h(d - N)] = b[1 + h(d - N)] = 0.$$

For the proof of the projectivity of P it remains to show that $1 + h(d - N)$ is not a right divisor of zero in kG.

Let $z \in kG$ and

$$z[1 + h(d - N)] = 0. \tag{11}$$

Then $z(1 + hd) = zfg \equiv 0 \bmod kGN$, whence from assumptions in the theorem we obtain $zf \in kGN$ and $z \in kGN$, so that $z = yN$ for some $y \in kG$. But then in (11) we have

$$0 = z[1 + h(d - N)] = yN[1 + h(d - N)] = yN = z.$$

Thus the module P is projective.

We consider the question of whether P is free. As in [17] we show that if P is free, then the image of f in $(k/d)(G/H)$ is the image of an invertible element in the ring kG/kGN. Let $P = kGz$, a principal left ideal. Then there exists an isomorphism of left kG-modules $\psi: P \to kG$. We note that by [14], Chapter 1, §1, $kGN = N(kG)$ consists of all $u \in kG$ such that $xu = u$ for all $x \in H$. Hence $\psi(N) = aN$, where a is an invertible element in $k(G/H)$. Let $\psi(f) = b \in kG$. Then from $Nf = fN$ we obtain

$$0 = N\psi(f) - f\psi(N) = Nb - faN,$$

i.e.,
$$b - fa \in \operatorname{Ann} N, \qquad f \equiv ba^{-1} \bmod \operatorname{Ann} N. \qquad (12)$$
We note that ψ induces an isomorphism of cyclic left kG-modules
$$P/kGN \to kG/kGN$$
with generators $f + kGN$, $1 + kGN$, respectively. Consequently b is invertible in kG/kGN, whence by (12) the element f is the image in $kG/(\operatorname{Ann} N + kGN)$ of the invertible element ba^{-1} in the ring kG/kGN. For the completion of the proof of the theorem it remains to note that
$$\operatorname{Ann} N = \sum_{x \in H} kG(x-1).$$
In fact,
$$\operatorname{Ann} N \cap kH = \sum_{x \in H} kH(x-1),$$
whence by Proposition 2 we obtain the required equality. The theorem is proved.

§5

In this section, using the preparatory work done above, we obtain a description of the finitely generated infinite nilpotent groups over whose integral group rings all projective modules are free.

THEOREM 4. *Let G be a finitely generated infinite nilpotent group, and let k be a commutative ring. Assume that one of the following conditions is satisfied:* 1) *If P is a left ideal in kG and $P \oplus kG \simeq kG \oplus kG$, then P is a principal left ideal.* 2) *If in the Noetherian domain k each prime number is noninvertible, then there exist a finite number of left ideals P in kG, nonisomorphic as modules, such that $P \oplus kG \simeq kG \oplus kG$. Then the group G' is finite.*

PROOF. Let H be a maximal abelian normal subgroup of G, and let $C(H)$ be the centralizer of H in G. By [9], Theorem 16.2.6, we have $C(H) = H$.

LEMMA 1. *The group H is infinite.*

PROOF. By [9], Theorem 16.2.6, the group G/H is imbeddable in $\operatorname{Aut} H$. Consequently, if H were finite, then it would be true that $|G| \leq |H| \cdot |\operatorname{Aut} H|$, so G would be finite, which contradicts the hypothesis of the theorem. The lemma is proved.

We denote by T the torsion subgroup of G, and by D the complete inverse image in G of the torsion subgroup of G/H. Then G/D is torsion-free, and D/H is finite.

LEMMA 2. $[H, D] \subseteq T \cap H$.

PROOF. Let $d = |D/H|$. For any $z \in D$ we have $[H, z^d] = 1$. We denote the ith term of the lower central series of D by D_i. By formula 33.34 of

[18], each group $[H, D_i]/[H, D_{i+1}]$ has exponent dividing d. Therefore if $D_{c+1} = 1$, then $[H, z]^{d^c} = 1$, so $[H, D] \subseteq T$. The lemma is proved.

LEMMA 3. $D' \subseteq T$.

PROOF. Let $d = |D/H|$. For each $z \in D$, we have $[D, z^d] \subseteq T$, by Lemma 2. But then, as in the proof of Lemma 2, we obtain the assertion of Lemma 3.

LEMMA 4. $[H, G] \subseteq T \cap H$.

PROOF. The group G/D is torsion-free. By [9], Theorem 17.2.2, there exists a central series $D = G_1 \subset G_2 \subset \cdots \subset G_r = G$ with infinite cyclic factors. We put $G_0 = T \cap D$. Then the factors of the series $G_0 \subset G_1 \subset \cdots \subset G_r = G$ are free abelian groups. Indeed, by Lemma 3 $G_1/G_0 \simeq D/(T \cap D)$ is a finitely generated torsion-free abelian group. The group H has finite index in D. Taking into account Lemma 1 and [8], Theorem 1, we get that for any $y \in H$ and $x \in G$, the elements y and y^x are related in G_1/G_0. Thus, conjugation by x gives in the free abelian group $H/(T \cap H) \subseteq G_1/G_0 = D/(T \cap D)$ the automorphism $y \to y^x$, where y and y^x are related. But then $y^x = y^\varepsilon$, where $\varepsilon = \varepsilon(x) = \pm 1$ is constant for all $y \in H$. If $y^x = y^{-1}$, then for any $c \geq 1$

$$[y, \overbrace{x, \ldots, x}^{c}] = y^{-2^c}.$$

Thus $\varepsilon(x) = 1$ for all $x \in G$, so that $[H, G] \subseteq T \cap H$. The lemma is proved.

It has already been noted that $C(H) = H$, so that G/H is imbedded in Aut H, where each $x \in G$ corresponds to conjugation by x. Since the group $[H, G]$ is finite, G/H is also finite, so that $G = D$. Hence, by Lemma 3, we get the assertion of the theorem.

THEOREM 5. *Let G be a finitely generated infinite nilpotent group such that all projective $\mathbb{Z}G$-modules are free. Then $G = A \times T$, where A is a free abelian group and T is a cyclic group of order 1, 2, 3, 5, 7, 11, 13, 17, or 19.*

PROOF. By Theorem 4, if T is the torsion subgroup of G, then G/T is a free abelian group. Let $t_1, \ldots, t_n \in G$ be a basis for G/T. As was noted in §§2 and 3, there exists $m \geq 1$ such that $f = t_1^m$ is in the center of G. Let $|T| = p_1^{n_1} \cdots p_r^{n_r}$, where the $n_i \geq 1$ and the p_i are prime numbers. Put $q = p_1 \cdots p_r$. We note that the element $1 + gf$ is invertible in the ring $\mathbb{Z}G/\mathbb{Z}G|T|$. Consequently, by Theorem 3 we have that $1 + qf$ is the image of an element in $\mathbb{Z}G$ that is invertible modulo $N = \sum_{x \in T} x$.

LEMMA 1. *Let the group T be noncyclic. Then there exist a prime number p dividing the order of T and a normal subgroup D of G such that $D \subseteq T$ and T/D is the direct product of two cyclic groups of order p.*

PROOF. The group T is the direct product of its Sylow subgroups. Hence with no loss of generality it may be assumed that T is a p-group. By [18],

Proposition 31.25, T/T' is noncyclic. Therefore the group T/T^pT' is a noncyclic abelian group of exponent p. It remains to use [9], Theorem 17.2.2.

Factoring by D we get that $1 + gf$ is the image of an element in $\mathbb{Z}\mathcal{U}$, where $\mathcal{U} = G/D$, that is invertible modulo $N_1 = \sum_{x \in T/D} x$. But this is impossible by Theorem 2, because $q = p_1 \cdots p_r$ is not divisible by any square. So the lemma is proved.

LEMMA 2. *The group T is cyclic.*

LEMMA 3. $|T| = q$.

Both lemmas follow from Theorem 1.

Thus, $|T| = q = p_1 \cdots p_r$. In particular, $T = \{x\}$ is contained in the center of the group G. We show that G is abelian. Suppose that $G' \neq 1$. Then $G' \subseteq T$. Recall that a basis $t_1, \ldots, t_n$ was chosen for the free abelian group G/T. Assume for example that $[t_1, t_1] = u \in T\setminus 1$. Then $t_1^{t_2} = t_1 u$ commutes with t_1. Therefore setting $y = t_1$ and $x = t_2$ in Propositions 1–3 of [8] we get that the left ideal in $\mathbb{Z}G$ generated by

$$h_1 = (t_1 u^{-1} - 1)(t_1 - 1)$$

and

$$h_2 = t_2(t_1 u^{-1} - 1) - (t_1 u^{-2} - 1)(t_1 - 1)$$

is a principal left ideal in $\mathbb{Z}GF$. Let the prime number p divide the order m of u. The element $\Phi_p(u)$ belongs to the center of $\mathbb{Z}G$. Therefore $I = \mathbb{Z}G\Phi_p(u) + \mathbb{Z}G(x^{mp^{-1}} - 1)$ is an ideal in $\mathbb{Z}G$. Using this ideal to form a quotient ring, we obtain α-extensions of rings ([8], §1)

$$\mathbb{Z}G/I \supset \mathbb{Z}G_1/(I \cap \mathbb{Z}G_1) \supset \mathbb{Z}T/(I \cap \mathbb{Z}T),$$

where G_1 is the subgroup generated by T and t_1. Consequently, by [8], Proposition 5, it can be assumed that, modulo I, F is a subset of $\mathbb{Z}G_1$. Hence by [8], Proposition 6, we obtain that

$$t_1 u^{-1} - 1, (t_1 u^{-2} - 1)(t_1 - 1) \in \mathbb{Z}G_1 F + (I \cap \mathbb{Z}G_1).$$

But then, similarly, $(u^{-1} - 1)(u - 1) \neq 0$ belongs to the ideal $\mathbb{Z}GF$. Thus, by [8], Proposition 5, we may assume that $F \subseteq \mathbb{Z}T + (I \cap \mathbb{Z}T)$. Then from the fact that $t_1 u^{-1} - 1 \in \mathbb{Z}G_1 F + (I \cap \mathbb{Z}G_1)$ we get that F is invertible in $\mathbb{Z}T/(I \cap \mathbb{Z}T)$. But this is impossible, since $I + \mathbb{Z}Gh_1 + \mathbb{Z}Gh_2$ is contained in the proper ideal $\mathbb{Z}Gp + \mathfrak{m}$, where $\mathfrak{m}$ is the fundamental ideal. This contradiction shows that G is an abelian group.

Thus $G = A \times T$, where A is a free abelian group and T is a cyclic group of order $q = p_1 \cdots p_r$. It is clear that under the hypotheses of the theorem the group $K_0(\mathbb{Z}G) \simeq \mathbb{Z}$ is trivial, so that by [19], p. 695, Theorem 10.6, the group T has prime order and also the class group $\text{Cl}(\mathbb{Z}T)$ is trivial. Therefore by [1] the order of T must be one of the numbers listed in the statement of the theorem. The theorem is proved.

THEOREM 6. *Let $G = A \times T$ be an abelian group, where A is a free abelian group of finite rank and T is a cyclic group of order* 1, 2, 3, 5, 7, 11, 13, 17, *or* 19. *Then all projective $\mathbb{Z}G$-modules are free.*

PROOF. By [19], p. 695, Theorem 10.6, $K_0(\mathbb{Z}G) \simeq \mathbb{Z}$. Therefore each finitely generated projective $\mathbb{Z}G$-module is stably free. Thus, it remains to show that the group $\mathrm{SL}(n, \mathbb{Z}G)$ acts on the right transitively on the set of all unimodular n-tuples, $n \geq 2$. Since the ring $\mathbb{Z}G$ is commutative, the case $n = 2$ is obvious. Let $n \geq 3$. If $p = |T|$, then for the listed values of p the ring $R_p = \mathbb{Z}[e^{2\pi i/p}]$ is a principal ideal ring [20]. By [21] the group $\mathrm{SL}(n, R_p A)$ acts transitively on the set of unimodular vectors. In addition, $\mathrm{SL}(n, R_p A) = E(n, R_p A)$. Therefore if we apply $E(n, \mathbb{Z}G)$ it can be assumed that the original unimodular vector has the form

$$(1 + a_1 \Phi_p(x), a_2 \Phi_p(x), \ldots, a_n \Phi_p(x)), \quad a_i \in \mathbb{Z}A,$$

where x is some generator of the group T. In this case the vector $(1 + pa_1, pa_2, \ldots, pa_n) \in (\mathbb{Z}A)^n$ is unimodular, so that there exists a matrix $1 + pB \in \mathrm{SL}(n, \mathbb{Z}A, p)$ such that

$$(1 + pa_1, pa_2, \ldots, pa_n)(1 + pB) = (1, 0, \ldots, 0),$$

or

$$(a_1, \ldots, a_n) + (1 + pa_1, pa_2, \ldots, pa_n)B = 0.$$

Hence

$$(1 + a_1 \Phi_p(x), a_2 \Phi_p(x), \ldots, a_n \Phi_p(x))(1 + \Phi_p(x)B)$$
$$= (1, 0, \ldots, 0) + \Phi_p(x)[(a_1, \ldots, a_n) + (1 + pa_1, pa_2, \ldots, pa_n)B]$$
$$= (1, 0, \ldots, 0).$$

In addition, if $(1 + pB)(1 + pC) = 1$ in $\mathrm{SL}(n, \mathbb{Z}A, p)$, then $(1 + \Phi_p(x)B) \cdot (1 + \Phi_p(x)C) = 1$ in $\mathrm{SL}(n, \mathbb{Z}G)$. The theorem is proved.

BIBLIOGRAPHY

1. Philippe Cassou-Noguès, *Classes d'idéaux d'algèbre d'un groupe abélian*, C. R. Acad. Sci. Paris Sér. A **276** (1973), 973–975.
2. Shizuo Endo and Yumiko Hironaka, *Finite groups with trivial class groups*, J. Math. Soc. Japan **31** (1979), 161–174.
3. Irving Reiner, *Class groups and Picard groups of group rings and orders*, Amer. Math. Soc., Providence, RI, 1976.
4. A. A. Suslin, *Projective modules over polynomial rings are free*, Dokl. Akad. Nauk SSSR **229** (1976), 1063–1066; English transl. in Soviet Math. Dokl. **17** (1976).
5. Richard G. Swan, *Projective modules over Laurent polynomial rings*, Trans. Amer. Math. Soc. 237 (1978), 111–120.
6. M. J. Dunwoody, *The homotopy type of a two-dimensional complex*, Bull. London Math. Soc. **8** (1976), 282–285.
7. P. H. Berridge and M. J. Dunwoody, *Nonfree projective modules for torsion-free groups*, J. London Math. Soc. (2) **19** (1979), 433–436.
8. V. A. Artamonov, *Projective nonfree modules over group rings of solvable groups*, Mat. Sb. **116** (1981), no. 2, 232–244; English transl. in Math. USSR-Sb. **44** (1983).

9. M. I. Kargapolov and Yu. I. Merzlyakov, *Fundamentals of the theory of groups*, "Nauka", Moscow, 1977; English transl., Springer-Verlag, Berlin and New York, 1979.
10. F. T. Farrell and W. C. Hsiang, *A formula for $K_1 R_\alpha[T]$*, Applications of Categorical Algebra, Proc. Sympos. Pure Math., vol. 17, Amer. Math. Soc., Providence, RI, 1970, pp. 192–218.
11. _____, *The Whitehead group of poly-(finite or cyclic) groups*, J. London Math. Soc. **24** (1981), 308–324.
12. K. A. Brown, T. H. Lenagan, and J. T. Stafford, *K-theory and stable structure of some Noetherian group rings*, Proc. London Math. Soc. (3) **42** (1981), 193–230.
13. J. Lewin, *Projective modules over group-algebras of one relator groups*, Abstracts Amer. Math. Soc. **1** (1980), 617.
14. A. E. Zalesskiĭ and A. V. Mikhalev, *Group rings*, Itogi Nauki i Tekhniki: Sovremennye Problemy Mat., vol. 2, VINITI, Moscow, 1973, pp. 5–118; English transl. in J. Soviet Math. **4** (1975), no. 1.
15. Robert B. Warfield, Jr., *Nilpotent groups*, Lecture Notes in Math., vol. 513, Springer-Verlag, Berlin and New York, 1976.
16. Z. I. Borevich and I. R. Shafarevich, *Theory of numbers*, "Nauka", Moscow, 1972; English transl. of 1st ed., Academic Press, New York and London, 1966.
17. Richard G. Swan, *Periodic resolutions for finite groups*, Ann. of Math. (2) **72** (1960), 267–291.
18. Hanna Neumann, *Varieties of groups*, Springer-Verlag, Berlin and New York, 1967.
19. H. Bass, *Algebraic K-theory*, Benjamin, New York, 1968.
20. J. Myron Masley and Hugh L. Montgomery, *Cyclotomic fields with unique factorization*, J. Reine Angew. Math. **286/287** (1976), 248–256.
21. A. A. Suslin, *On the structure of the special linear group over polynomial rings*, Izv. Akad. Nauk SSSR Ser. Mat. **41** (1977), no. 2, 235–252; English transl. in Math. USSR-Izv. **11** (1977).

Translated by C. W. KOHLS

On Torsion in Higher Milnor Functors for Multidimensional Local Fields

S. V. VOSTOKOV AND I. B. FESENKO

Let k be a local field, and let m be the number of roots of unity in k. A well-known theorem of C. Moore asserts that $mK_2(k)$ is a divisible subgroup of the group $K_2(k)$, and the corresponding quotient group is isomorphic to a cyclic group of order m. J. Carroll [6] and A. S. Merkur'ev [3] showed that the group $mK_2(k)$ has no torsion. I. Ya. Sivitskiĭ used Hilbert's Theorem 90 for K_2 to prove that the higher K-groups for the field k are uniquely divisible.

Meanwhile, K. Kato [7], [8] and A. N. Parshin [4], [5] constructed a generalization of local class field theory to multidimensional local fields. Interest in such fields (as the natural generalization of a classical local field) arose first of all in connection with their role in algebraic geometry; in addition, the study of questions connected with these fields entails the further development of local class field theory itself.

In the present paper we investigate the groups $K_{n+i}(F)$ and $K^*_{n+i}(F)$, where F is an n-dimensional local field, i is a positive integer, and $K^*_{n+i}(F)$ is the separable quotient group of $K_{n+i}(F)$ relative to a certain topology (for the definition, see §1). We obtain results for these groups which are analogous to some of the results mentioned above for one-dimensional local fields.

§1. Moore's theorem for multidimensional local fields

1°. A field F is said to be an n-dimensional local field if there exists a chain of complete discrete valuation fields $k_0, \ldots, k_n = F$ which satisfy the following conditions: $k_0 = \mathbb{F}_q$ is a finite field, and k_{i-1} is the residue field of k_i. By lifting prime elements of the fields k_i to F, we obtain a system of parameters $t_1, t_2, \ldots, t_n$ for F, and this determines a rank n valuation of the field F [2]: $v\colon F^* \to \mathbb{Z}^{\oplus n}$ (if A is a group, then we let $A^{\oplus n}$ denote the direct sum of n copies of A), $v(t_m) = (0, \ldots, 1, \ldots, 0)$, where the one is at the $(n-m)$th place. We introduce the following notation for the ideals of the ring of integers $\mathscr{O}_F$ of the field F relative to the valuation v:

$$\mathfrak{P}_v(i_1, \ldots, i_n) = \{x \in F \mid v(x) \geq (i_1, \ldots, i_n)\},$$

$$\mathfrak{P}_v^l(i_1, \ldots, i_l) = \bigcup_{i \in \mathbb{Z}} \mathfrak{P}_v^{l+1}(i_1, \ldots, i_l, i).$$

Let

$$\mathfrak{M}_F = \mathfrak{P}_v(0, 0, \ldots, 1)$$

and let U_F be the group of units of $\mathscr{O}_F$. The group $(\mathscr{O}_F/\mathfrak{M}_F)^*$ is isomorphic to the multiplicative group $\mathfrak{R}$ of roots of unity in F of order prime to p, where p is the characteristic of the first residue field k_0.

According to Parshin [4], there are three types of n-dimensional local fields: either $F = \mathbb{F}_q((t_1))\cdots((t_n))$, or else $F = k((t_2))\cdots((t_n))$, or else F is a finite extension of the field $k\{\{t_1\}\}\cdots\{\{t_l\}\}((t_{l+2}))\cdots((t_n))$, where k is a local number field of dimension one, and it is contained in a field of the same type (possibly with different k and t_i).

We define the groups $K_m^*(F)$ for a field which is a finite extension of fields of the form $k\{\{\tilde{t}_1\}\}\cdots\{\{\tilde{t}_{n-1}\}\}$. Let $t_1, \ldots, t_n$ be a system of parameters for F. As a set of representatives modulo $\mathfrak{P}_v^1(1)$ in $\mathscr{O}_F$ we take the set

$$S = \{0\} \cup \mathfrak{R}((t_1))\cdots((t_{n-1})).$$

In F^* we introduce the topology which is the product of the multiplicative topology on k_{n-1}^*, the discrete topology on $\mathbb{Z}$, and the following topology on $1 + \mathfrak{P}_v^1(1)$: the neighborhoods of one are the sets of elements of the form $1 + t_n u_1 + t_n^2 u_2 + \cdots$, where the u_i are in the subsets of S which give neighborhoods of zero in the additive topology of k_{n-1}. On $K_m(F)$ we now take the strongest topology in which the map $\varphi\colon F^{*\oplus m} \mapsto K_m(F)$ is continuous in each argument and $K_m(F)$ is a topological group. We obtain $K_m^*(F)$ by taking the quotient of $K_m(F)$ by the subgroup which is the intersection of all of the neighborhoods of zero. We note that, if we require that φ be continuous in the product of the multiplicative topologies on $F^{*\oplus m}$, then we obtain the groups $K_m^{\text{top}}(F)$, which were introduced by Parshin and which play an important role in the generalization of local class field theory. For one-dimensional local fields one has $K_m^*(F) = K_m(F)$. If $x = \{d_1, \ldots, d_m\} \in K_m(F)$, then we shall use the same symbol x for the image of this element in $K_m^*(F)$.

2°. We define $v_i = p_i \circ v$, where $p_i: \mathbb{Z}^{\oplus n} \mapsto \mathbb{Z}$ is the projection onto the ith component. We consider the matrix
$$A = \begin{pmatrix} v_1(\alpha_1) & v_1(\alpha_2) & \cdots & v_1(\alpha_{n+1}) \\ v_2(\alpha_1) & v_2(\alpha_2) & \cdots & v_2(\alpha_{n+1}) \\ \cdots & \cdots & \cdots & \cdots \\ v_n(\alpha_1) & v_n(\alpha_2) & \cdots & v_n(\alpha_{n+1}) \end{pmatrix}.$$
We let A_i denote the determinant of the matrix which is obtained from A by crossing out the ith column, multiplied by $(-1)^{i-1}$; we let A_{ij}^I denote the determinant of the matrix obtained from A by crossing out the ith and jth columns and the Ith row.

We define the map
$$c_{\text{tame}}^{(n)}: F^{*\oplus(n+1)} \mapsto (\mathscr{O}_F/\mathfrak{M}_F)^* \cong \mathfrak{R}$$
by the formula $c_{\text{tame}}^{(n)}(\alpha_1, \ldots, \alpha_{n+1}) = \alpha_1^{A_1} \alpha_2^{A_2} \cdots \alpha_{n+1}^{A_{n+1}} (-1)^B \mod \mathfrak{M}_F$, where $B = \sum_{i<j, I} v_I(\alpha_i) v_I(\alpha_j) A_{ij}^I$.

PROPOSITION 1. *The map $c_{\text{tame}}^{(n)}$ is well defined, and it is the Steinberg symbol, with $c_{\text{tame}}^{(n)}(\theta, t_n, t_{n-1}, \ldots, t_1) = \theta$, where $\theta \in \mathfrak{R}$.*

PROOF. The skew symmetry, linearity, and correctness are easily verified. We show that $c_{\text{tame}}^{(n)}(\alpha, 1-\alpha, \alpha_3, \ldots, \alpha_{n+1}) = 1$, if $\alpha \ne 0, 1$. Let s be the greatest number such that $v_1(\alpha) = \cdots = v_s(\alpha) = 0$, $v_1(1-\alpha) = \cdots = v_s(1-\alpha) = 0$, and $v_{s+1}(\alpha)$ or $v_{s+1}(1-\alpha)$ is nonzero. If $v_{s+1}(\alpha) > 0$, then $v_{s+1}(1-\alpha) = \cdots = v_n(1-\alpha) = 0$ and $c_{\text{tame}}^{(n)}(\alpha, 1-\alpha, \alpha_3, \ldots, \alpha_{n+1}) \equiv 1 \mod \mathfrak{P}_v^{n-s}(0, \ldots, 1)$. If $v_{s+1}(\alpha) < 0$, then $v_{s+1}(\alpha) = v_{s+1}(1-\alpha), \ldots, v_n(\alpha) = v_n(1-\alpha)$, and hence $A_1 = -A_2$, $A_3 = \cdots = A_{n+1} = 0$ and
$$B = \sum_I v_I(\alpha) v_I(1-\alpha) A_{1,2}^I \equiv \sum_I v_I(\alpha) A_{1,2}^I \equiv A_1 \mod \mathbb{Z};$$
hence,
$$c_{\text{tame}}^{(n)}(\alpha, 1-\alpha, \alpha_3, \ldots, \alpha_{n+1}) = \alpha^{A_1} (1-\alpha)^{-A_1} (-1)^{A_1}$$
$$= (1-\alpha^{-1})^{-A_1} \equiv 1 \mod \mathfrak{P}_v^{n-s}(0, \ldots, 1).$$
The proposition is proved.

REMARK. The tame symbol for a multidimensional local field was first introduced (up to sign) by Parshin (see Chapter 3 of [5]).

3°. LEMMA 1. *The groups*
$$K_{n+1}(F)/(q-1)K_{n+1}(F) \quad \text{and} \quad K_{n+1}^*(F)/(q-1)K_{n+1}^*(F)$$
are cyclic of order $q-1$, and are generated by $\{t_1, \ldots, t_n, \theta\}$, $\theta \in \mathfrak{R}$.

PROOF. If $\theta \in \mathfrak{R}$ and $u \in U_F$, then $\{\theta, u\} = 0$, and hence the group $K_{n+1}(F)/(q-1)K_{n+1}(F)$ is generated by $\{t_1, \ldots, t_n, \theta\}$. It follows from the existence of the tame symbol that the order of $K_{n+1}(F)/(q-1)K_{n+1}(F)$ is not less than $q-1$. The lemma is proved.

Let F be a finite extension of $k\{\{\tilde{t}_1\}\}\cdots\{\{\tilde{t}_{n-1}\}\}$, where k is a one-dimensional local number field. Let
$$p \equiv \theta t_n^{e_n} t_{n-1}^{e_{n-1}} \cdots t_1^{e_1} \bmod \mathfrak{P}_v(e_n, e_{n-1}, \ldots, e_1 + 1),$$
where $t_1, \ldots, t_{n-1}, t_n$ is a system of parameters for the field F; and set $e_i' = e_i/(p-1)$. We choose the following generators of the group of principal units $(1 + \mathfrak{M}_F)^*$ over $\mathbb{Z}_p$: first of all, the elements of the form $(1 \pm \theta t_n^{u_n} t_{n-1}^{u_{n-1}} \cdots t_1^{u_1})^a$, where $(0, \ldots, 0) < (u_n, u_{n-1}, \ldots, u_1) < (pe_n', \ldots, pe_1')$ and g.c.d. $(u_1, u_2, \ldots, u_n, p) = 1$, the integer a is any of the u_i which is not divisible by p, θ is from the basis of $\mathfrak{R}$ over $\mathbb{F}_p$, and the sign is chosen so that $\{t_n, t_{n-1}, \ldots, t_1, (1 \pm \theta t_n^{u_n} t_{n-1}^{u_{n-1}} \cdots t_1^{u_1})^a\} = 0$. In addition, if $\xi_p \in F$, then the set of generators over $\mathbb{Z}_p$ of the group of principal units must also include the element $\delta = (1 - \eta t_n^{pe_n'} \cdots t_1^{pe_1'})^b$, where $1 - \eta t_n^{pe_n'} \cdots t_1^{pe_1'} \notin F^{*p}$ and b is the largest divisor of $p^n e_1' \cdots e_n'$ which is not divisible by p.

LEMMA 2. *If $u \in (1 + \mathfrak{M}_F)^*$, then $\{u, \delta\} \equiv 0 \bmod pK_2(F)$.*

PROOF. We find an element θ in $\mathfrak{R}$ such that $\theta^p = \eta$; then, since $1 + \mathfrak{P}_v(pe_n', \ldots, pe_1' + 1) \subset F^{*p}$, there exists $w \in U_F$ such that $w^p = 1 - \eta t_n^{pe_n'} \cdots t_1^{pe_1'} + (\theta t_n^{e_n'} \cdots t_1^{e_1'})^p u$. From this we conclude that $\{1 - \eta t_n^{pe_n'} \cdots t_1^{pe_1'}, u\} \equiv 0 \bmod pK_2(F)$.

A map $c: F^{*\oplus m} \mapsto A$, where A is a topological group, will be said to be continuous if it is continuous when $F^{*\oplus m}$ is given the direct product topology coming from the multiplicative topology on F^*. If c is a continuous map relative to the product of the topologies on F^* given in $1°$, and if $rc = 0$ for some natural number r, then c is continuous, since there exists i for which $1 + \mathfrak{P}_v^1(i) \subset F^{*m}$.

S. V. Vostokov [1] constructed the Hilbert symbol for fields of the third type. If ξ_{p^t} is a primitive p^tth root of unity in F, then there exists a continuous map
$$\omega_{p^t}: K_{n+1}(F) \mapsto \mu_{p^t} = (\xi_{p^t}),$$
having the norm property and the nondegeneracy property.

LEMMA 3. *Suppose that F contains a primitive p^tth root of unity. Then for any $r \leq t$ the group $K_{n+1}^*(F)/p^r K_{n+1}^*(F)$ is a cyclic group of order p^r, and it is generated by $\{t_1, t_2, \ldots, t_n, \delta\}$.*

PROOF. We consider the symbol mapping
$$c: F^{*\oplus(n+1)} \mapsto K_{n+1}^*(F)/pK_{n+1}^*(F).$$
By the remark above, c is continuous. Hence, by Lemma 2, the group $K_{n+1}^*(F)/pK_{n+1}^*(F)$ is generated by $\{t_1, t_2, \ldots, t_n, \delta_{t_1', \ldots, t_n'}\}$, where $t_1', \ldots, t_n'$ is a certain system of parameters for F. Expanding $\delta_{t_1', \ldots, t_n'}$ in the generators of the group of principal units over $\mathbb{Z}_p$, we find that the group

$K_{n+1}^*(F)/pK_{n+1}^*(F)$ is generated by $\{t_1, \ldots, t_n, \delta\}$. The lemma now follows from the existence of the Hilbert symbol of degree p^r.

LEMMA 4. *Let m be the number of roots of unity in the field F. Then $K_{n+1}^*(F)/mK_{n+1}^*(F)$ is a cyclic group of order m.*

PROOF. Let ξ_m be a primitive mth root of unity in F. Then, by the nondegeneracy of the Hilbert symbol of degree p, there exist $\alpha_1, \ldots, \alpha_n \in F^*$, such that $\omega_p(\{\alpha_1, \ldots, \alpha_n, \xi_m\}) = \xi_p$; hence, there exists a system of parameters $t_1', \ldots, t_n'$, such that $\{t_1', \ldots, t_n', \xi_m\}$ generates $K_{n+1}^*(F)/pK_{n+1}^*(F)$, but $\{t_1', \ldots, t_n', \xi_m^{m/(q-1)}\}$ generates $K_{n+1}^*(F)/(q-1)K_{n+1}^*(F)$, so that $K_{n+1}^*(F)/mK_{n+1}^*(F)$ is a cyclic group of order m.

THEOREM 1. *If m is the number of roots of unity in F, then the group $K_{n+1}^*(F)$ decomposes as the direct sum of a cyclic group of order m and the divisible subgroup $mK_{n+1}^*(F)$.*

PROOF. The fact that $mK_{n+1}^*(F)$ is a divisible group follows from the continuity of the map $c: F^{*\oplus(n+1)} \mapsto K_{n+1}^*(F)/mrK_{n+1}^*(F)$, where r is a positive integer.

REMARK 1. The continuous Hilbert symbol of degree p^t, where $t = \mathrm{ord}_p m$, is the universal wild symbol; a formula for it is given in [1]. The tame symbol in 1° is the universal tame symbol.

REMARK 2. From Theorem 3 and the corollary in [8], p. 672, it follows that $K_{n+1}(F)/mK_{n+1}(F)$ is a cyclic group of order m.

§2. Prime to p torsion in the groups $K_{n+i}(F)$ and $K_{n+i}^*(F)$

Let F be an n-dimensional local field. We fix a prime number l different from p, and we let C be the group of lth power roots of unity of degree prime to p which are contained in F. Let M be the subgroup of $K_{n+1}(F)$ consisting of elements of the form $\{\alpha, \beta_1, \ldots, \beta_n\}$, where $\alpha \in C$ and $\beta_i \in F^*$.

PROPOSITION 2. *If $b \in C$, $w \in V$, and $x \in K_{n-1}(F)$, then $\{1-bw^l, w\} \cdot x = 0$ in $K_{n+1}(F)$.*

The proof is similar to J. Carroll's argument in [6].

We define the map $c_l: F^{*\oplus(n+1)} \mapsto K_{n+1}(F)$ as follows: let $\alpha_i = t_1^{m_{1,i}} \cdot t_2^{m_{2,i}} \cdots t_n^{m_{n,i}} b_i w_i$, where $b \in C$ and $w_i \in V$. We decompose the element $\{\alpha_1, \alpha_2, \ldots, \alpha_{n+1}\}$ in the natural way as a sum of terms of the form $a\{\beta_1, \beta_2, \ldots, \beta_{n+1}\}$, where β_i either is a prime parameter of F or else is in the set $C \cup V$, and $a \in \mathbb{Z}$. In the terms where $\beta_i = \beta_j$ is a prime parameter, we make a substitution using $\{\beta_i, \beta_j\} = \{\beta_i, -1\}$, until all of the β_i which are prime parameters and occur in a term of this type become distinct. Next, we discard all of the terms in which at least one of the β_i

belongs to C (where we take into account that $-1 \in C$ for $l = 2$). Finally, the remaining terms are of the form $a\{\gamma_1, \ldots, \gamma_{n+1}\}$, where at least one of the γ_i belongs to V. We extract the lth root of any one of the γ_i. The resulting element $K_{n+1}(F)$ is, by definition, $c_l(\alpha_1, \ldots, \alpha_{n+1})$.

For $n = 1$ it is easy to write an explicit formula for the map c_l:

$$c_l(t^{m_1} b_1 w_1, t^{m_2} b_2 w_2) = \{\pi, (-1)^{lm_1 m_2}\} + \{\pi, (w_2^{m_1} w_1^{-m_2})^{1/l}\} + \{w_1, w_2^{1/l}\}.$$

Note that c_l is well defined, since $\{w_1^{1/l}, w_2\} = l\{w_1^{1/l}, w_2^{1/l}\} = \{w_1, w_2^{1/l}\}$; in addition, c_l is skew symmetric and multiplicative. Hence,

$$c_l(1 - \alpha, \alpha, \alpha_3, \ldots, \alpha_{n+1}) + c_l(1 - \alpha^{-1}, \alpha^{-1}, \alpha_3, \ldots, \alpha_{n+1})$$
$$= c_l(-\alpha, \alpha, \alpha_3, \ldots, \alpha_{n+1}).$$

If we take into account that for every $\alpha \in F^*$ at least one of the three elements α, α^{-1}, $1 - \alpha$ belongs to $U_F = CV$, we see that to prove that c_l is a symbol it suffices to verify that

$$c_l(1 - bw, bw, \alpha_3, \ldots, \alpha_{n+1}) = 0 \quad \text{for all } b \in C, \; w \in V,$$
$$c_l(-\alpha, \alpha, \alpha_3, \ldots, \alpha_{n+1}) = 0 \quad \text{for all } \alpha \in F^*.$$

To prove the first equality we use the fact that

$$c_l(1 - bw, bw, \alpha_3, \ldots, \alpha_{n+1}) = 0$$

and

$$c_l(1 - bw, w, \alpha_3, \ldots, \alpha_{n+1}) = \{1 - bw, w^{1/l}, \alpha_3, \ldots, \alpha_{n+1}\} = 0,$$

by Proposition 2. We prove the second equality in two stages. If l is odd and $\alpha = t_1^{m_1} t_2^{m_2} \cdots t_n^{m_n} bw$, then $-\alpha = t_1^{m_1} \cdots t_n^{m_n} b(-w)$. Hence,

$$c_l(-\alpha, \alpha, \alpha_3, \ldots, \alpha_{n+1})$$
$$= c_l(t_1^{m_1} t_2^{m_2} \cdots t_n^{m_n}, t_1^{m_1} t_2^{m_2} \cdots t_n^{m_n}, \ldots) + c_l(t_1^{m_1} t_2^{m_2} \cdots t_n^{m_n}, -w, \ldots)$$
$$+ c_l(w, t_1^{m_1} \cdots t_n^{m_n}, \ldots) + c_l(w, -w, \ldots)$$
$$= \{t_1^{m_1} \cdots t_n^{m_n}, -1, \ldots\} + \{t_1^{m_1} \cdots t_n^{m_n}, (-w)^{1/l}, \ldots\}$$
$$+ \{w^{1/l}, t_1^{m_1} \cdots t_n^{m_n}, \ldots\} + \{w, (-w)^{1/l}, \ldots\}.$$

But $(-w)^{1/l} = -w^{1/l}$, and so

$$c_l(-\alpha, \alpha, \alpha_3, \ldots, \alpha_{n+1}) = 2\{t_1^{m_1} \cdots t_n^{m_n}, -1, \ldots\} = 0.$$

If l is even and $\alpha = t_1^{m_1} \cdots t_n^{m_n} bw$, then $-\alpha = t_1^{m_1} \cdots t_n^{m_n}(-b)w$. In this case we have

$$c_l(-\alpha, \alpha, \alpha_3, \ldots, \alpha_{n+1}) = \{t_1^{m-1} \cdots t_n^{m_n}, w^{1/l}, \ldots\}$$
$$+ \{w^{1/l}, t_1^{m_1} \cdots t_n^{m_n}, \ldots\} + \{w^{1/l}, w^{1/l}, \ldots\},$$

since $c_l(t_1^{m_1}\cdots t_n^{m_n}, t_1^{m_1}\cdots t_n^{m_n}, \ldots) = 0$ and $\{w^{1/l}, w\} = 2\{w^{1/2}, -1\} = 0$. Thus, the map c_l factors through $K_{n+1}(F)$, and for $x \in K_{n+1}(F)$ we have

$$c_l(lx) \equiv x \bmod M.$$

This immediately implies that if $lx = 0$, then $x \in M$.

THEOREM 2. *If ξ is a primitive lth root of unity in F, $(l, p) = 1$, and $x \in K_{n+i+1}(F)$, $l \geq 0$, is an element of order a power of l, then $x = \{\xi\}y$, where $y \in K_{n+i}(F)$.*

PROOF. If $i = 0$, then the theorem has already been proved. The proof is similar in the general case.

PROPOSITION 3. *The groups $(q-1)K_{n+1}(F)$, $(q-1)K_{n+1}^*(F)$, $K_{n+i+1}(F)$, and $K_{n+i+1}^*(F)$, where $i \geq 1$, have no prime to p torsion.*

PROOF. Note that if $x \in M$, then the equality $c_{\text{tame}}^{(n)}(x) = 1$ implies that $x = 0$ in $K_{n+1}(F)$. Hence, the claim of the proposition for the groups $(q-1)K_{n+1}(F)$ and $K_{n+i+1}(F)$ follows from Theorem 2. Furthermore, the intersection of the neighborhoods of 0 in $K_{n+i+1}(F)$ in the definition of $K_{n+i}^*(F)$ in §1 is $(q-1)$-divisible; hence, there is no l-torsion in the groups $K_{n+i+1}^*(F)$ and $(q-1)K_{n+1}^*(F)$.

At the end of this section we shall clarify the question of divisibility of the higher K^*-groups. At that point we shall make use of Theorem 3 of §3; however, the result obtained will not be used in the proof of the latter.

PROPOSITION 4. *The groups $K_{n+i+1}^*(F)$, where $i \geq 1$, are divisible.*

PROOF. Let $\{\alpha_1, \ldots, \alpha_{n+2}\} \in K_{n+2}^*(F)$. Applying Lemma 4 of §1, we write $\{\alpha_1, \ldots, \alpha_{n+1}\} = \{\xi_m, \beta_2, \ldots, \beta_{n+1}\} + x$, where $x \in mK_{n+1}^*(F)$. We next show that $\{\xi_m, \beta_2, \ldots, \beta_{n+1}, \alpha_{n+2}\} = 0$, and hence that $K_{n+2}^*(F)$ is divisible, since $mK_{n+1}^*(F)$ is a divisible group.

Applying Lemma 4 of §1 once more, we obtain $\{\xi_m, \beta_2, \ldots, \beta_{n+1}, \alpha_{n+2}\} = \{\xi_m, \xi_m, \gamma_3, \ldots, \gamma_{n+2}\}$. It is well known [6] that $\{\xi_m, \xi_m\} = 0$ if m is odd or if $4|m$. Otherwise, we have $\{\xi_m, \xi_m\} = a\{-1, -1\}$, a an integer. If $p \neq 2$, then $-1 \in \mathfrak{R}$, and so $\{-1, -1\} = 0$. If $p = 2$ and $\text{ord}_2 m = 1$, then we consider two cases. If $\{-1, -1, \gamma_3, \ldots, \gamma_{n+1}\} \in 2K_{n+1}^*(F)$, then it follows from Theorem 3 of §3 that $\{-1, -1, \gamma_3, \ldots, \gamma_{n+1}\} = 0$. Otherwise, by Theorem 1 of §1, $\{-1, -1, \gamma_3, \ldots, \gamma_{n+1}\}$ generates the group $K_{n+1}^*(F)/2K_{n+1}^*(F)$, and hence

$$\{-1, -1, \gamma_3, \ldots, \gamma_{n+2}\} = \{-1, -1, -1, \gamma_4', \ldots, \gamma_{n+2}'\}.$$

But $\{-1, -1, -1\} = 0$ in $K_3(\mathbb{Q}_2)$, and the proposition is proved.

§3. p-torsion

An inclusion of fields $i: F \mapsto L$ induces a map $l^*: K_m^*(F) \mapsto K_m^*(L)$, which is compatible with the corresponding map for K-groups.

QUESTION. Suppose that the group F^* is given a topology and $K_m^*(F)$ is given the strongest topology for which the map $F^{*\oplus m} \mapsto K_m(F)$ is continuous in each argument and $K_m(F)$ is a topological group. Let $\Lambda(F)$ be the intersection of the neighborhoods of 0 in $K_m(F)$. Under what conditions on the topology on F^* do we have a map $N: K_m(L)/\Lambda(L) \mapsto K_m(F)/\Lambda(F)$ defined for finite field extensions L/F which is compatible with the transfer map of K-theory?

In the case of n-dimensional local fields of the first type and the groups $K_n^{\text{top}}(F)$, the existence of the norm map was proved by Parshin in [5], using the Kummer theory and Artin-Schreier theory of multidimensional local fields. A positive answer to the above question for the groups $K_m^*(F)$ would make it possible to obtain results on the absence of p-torsion in these groups.

Next, to prove the analogue of Merkur'ev's theorem [3] we shall need the following lemmas.

LEMMA 5. *Let $g_1, g_2 \in K_n^*(F)$ and $z_1, z_2 \in F^*$. Then there exists an element x of the field F such that*

$$\omega_{p^t}(g_1 \cdot z_1) = \omega_{p^t}(g_1 \cdot x)^i, \qquad \omega_{p^t}(g_2 \cdot z_2) = \omega_{p^t}(g_2 \cdot x)^j,$$

where i and j are integers.

PROOF. We fix a primitive p^tth root of unity ξ_{p^t}. We define the map $\varphi_g: F^* \mapsto \mathbb{Z}/p^t\mathbb{Z}$ by the formula $\varphi_g(x) = i$, where $\omega_{p^t}(g \cdot x) = \xi_{p^t}^i$. Let l_1 and l_2 be the largest natural numbers such that the image of φ_{g_1} is divisible by p^{l_1} and the image of φ_{g_2} is divisible by p^{l_2}. By the nondegeneracy of the Hilbert symbol, there exist $\tilde{g}_1$ and $\tilde{g}_2$ such that $g_1 = p^{l_1}\tilde{g}_1$ and $g_2 = p^{l_2}\tilde{g}_2$. If there exists l for which the image of $\varphi_{\tilde{g}_1} - l\varphi_{\tilde{g}_2}$ is divisible by p, then, taking x so that $\varphi_{\tilde{g}_1}(x) \not\equiv 0 \bmod p$, we find that we also have $\varphi_{\tilde{g}_2}(x) \not\equiv 0 \bmod p$, and hence the required i and j exist. On the other hand, if $\varphi_{\tilde{g}_1}$ and $\varphi_{\tilde{g}_2}$ are linearly independent modulo p, then there exist $x_1, x_2 \in F^*$ such that

$$\varphi_{\tilde{g}_1}(x_1)\varphi_{\tilde{g}_2}(x_2) - \varphi_{\tilde{g}_1}(x_2)\varphi_{\tilde{g}_2}(x_1) \not\equiv 0 \bmod p.$$

Then x can be found in the form $x_1^a x_2^b$.

LEMMA 6. *Let $x = \{\xi_{p^t}\} \cdot y$, where $x \in mK_{n+1}^*(F)$. Then there exist $x_i \in mK_{n+1}^*(F)$ of the form $\{\xi_{p^t}, \beta_2, \ldots, \beta_{n+1}\}$, such that $x = \sum_{i=1}^N x_i$.*

PROOF. We represent x as a sum of N terms of the form $\{\xi_{p^t}, \alpha_2, \ldots, \alpha_{n+1}\}$. We prove the lemma by induction on the number N of terms. If $x = \{\xi_{p^t}, \alpha_2, \ldots, \alpha_{n+1}\} + \{\xi_{p^t}, \alpha_2', \ldots, \alpha_{n+1}'\}$, then, by Lemma 5, there exists $\gamma_1 \in F^*$ such that

$$\omega_{p^s}(\{\xi_{p^t}, \alpha_2, \ldots, \alpha_{n+1}\}) = \omega_{p^s}(\{\xi_{p^t}, \alpha_2, \ldots, \alpha_n, \gamma_1\})^{i_1},$$
$$\omega_{p^s}(\{\xi_{p^t}, \alpha'_2, \ldots, \alpha'_{n+1}\}) = \omega_{p^s}(\{\xi_{p^t}, \alpha'_1, \ldots, \alpha'_n, \gamma_1\})^{i_2},$$

where $s = \operatorname{ord}_p m$. Hence,

$$x = \{\xi_{p^t}, \alpha_2, \ldots, \alpha_n^{i_1}, \gamma_1\} + \{\xi_{p^t}, \alpha'_2, \ldots, \alpha'^{i_2}_n, \gamma_1\}$$
$$- \{\xi_{p^t}, \alpha_2, \ldots, \alpha_n, \gamma_1^{i_1}\alpha_{n+1}^{-1}\} - \{\xi_{p^t}, \alpha'_2, \ldots, \alpha'_n, \gamma_1^{i_2}\alpha'^{-1}_{n+1}\}.$$

The last two terms belong to $mK^*_{n+1}(F)$, and the first two have two elements in common: ξ_{p^t} and γ_1. If we repeat this procedure $n-1$ times, we find that

$$x = j_1\{\xi_{p^t}, \gamma_n, \ldots, \gamma_1\} + j_2\{\xi_{p^t}, \gamma_n, \ldots, \gamma_1\} + x'.$$

where $x' \in mK^*_{n+1}(F)$, and x satisfies the requirements of the lemma. But $(j_1 + j_2)\{\xi_{p^t}, \gamma_n, \ldots, \gamma_1\} \in mK^*_{n+1}(F)$.

The induction step. Let $x = \sum_{i=1}^N x_i$, $N \geq 3$. Suppose that $\omega_{p^s}(x_1)$ generates the set consisting of $\omega_{p^s}(x_1), \ldots, \omega_{p^s}(x_N)$. We find $l \in \mathbb{Z}$ such that $(l+1)x_1 + x_2 \in mK^*_{n+1}(F)$ and we write $x = ((l+1)x_1 + x_2) + \sum_{i=3}^N x_i - lx_i$. Each term is in $mK^*_{n+1}(F)$, and we can apply the induction assumption. The lemma is proved.

LEMMA 7. *The group $K^*_n(F)/p^l K^*_n(F)$ is generated by the elements of the form $\{t_1, t_2, \ldots, t_n\}$, $\{t_1, \ldots, \hat{t}_k, \ldots, t_n, u\}$, where $u \in (1 + \mathfrak{M}_F)^*$ and $t_1, \ldots, t_n$ is a fixed system of parameters for the field F.*

PROOF. We use Proposition 1 of [5] and the fact that the map $F^{*\oplus n} \mapsto K^*_n(F)/p^l$ is continuous.

We suppose that the following holds: if $\xi_{p^r} \in F$ and $x \in K^*_{n+1}(F)$ is an element of order p^r, then there exists $y \in K^*_n(F)$ such that $x = \{\xi_{p^r}\} \cdot y$.

THEOREM 3. *The group $mK^*_{n+1}(F)$ has no p-torsion.*

PROOF. We shall assume that $F = k\{\{t_1\}\}\cdots\{\{t_{n_1}\}\}$, where k is a local number field of dimension 1. Suppose that e is the ramification index of $k/\mathbb{Q}_p$, $e' = e/(p-1)$, t is the largest number such that $\xi_{p^t} \in k$, $s = \operatorname{ord}_p(pe')$, and π is a prime element of the field k. We suppose that the field k satisfies the following three conditions:
1) the extension $k(\xi_{p^r})/k$ is totally ramified for any $r \geq 0$,
2) $2t \geq s + 1$ for the field k,
3) $\xi_{p^t} \notin U^p U_{e+1}$, where U is the group of units of the ring of integers of k, and $U_i = \{x \in k^* | x \equiv 1 \bmod \pi^i\}$.

In order to prove the theorem, we show that if $\{\xi_{p^t}, \alpha_2, \ldots, \alpha_{n+1}\} \in mK^*_{n+1}(F)$ is of order p^l, then $\{\xi_p, \alpha_2, \ldots, \alpha_{n+1}\} = 0$. In view of Lemma

6, it will then follow that $mK_{n+1}^*(F)$ has no p-torsion. We divide the proof into two parts.

1. We first let $\{\alpha_2, \ldots, \alpha_{n+1}\} = \{t_n, t_{n-1}, \ldots, t_1\}$. We expand ξ_{p^t} with respect to the generators in §1 in two ways:

$$\xi_{p^t} = (1 - \eta \pi^{pe'})^a \cdot \prod_{\theta, i} (1 - \theta \pi^i)^{a_{\theta, i}},$$

$$\xi_{p^t} = (1 - \eta t_n^{pe'_n})^b \prod_{\alpha = (i_1, \ldots, i_n)} (1 - \theta t_n^{i_n} \cdots t_1^{i_1})^{b_{\theta, \alpha}}.$$

From the equality $\omega_{p^t}(\{\xi_{p^t}, t_n, t_{n-1}, \ldots, t_1\}) = 1$ it follows that $b \equiv 0 \mod p^t$. The proof of Lemma 3 in [3] shows that there exists a smallest $i \leq e$ such that $a_{\theta, i}$ is not divisible by p. Since $\pi_0 = t_n u$, where $u \in U_F$, it follows that $(i, 0, \ldots, 0) \leq (e, 0, \ldots, 0)$ is the smallest for which $b_{\theta, (i, 0, \ldots, 0)}$ is not divisible by p. Hence, as in Lemma 3 of [3], one shows that

$$\Delta_{t_n} = \frac{pe'_n b \eta t_n^{pe'_n}}{1 - \eta t_n^{pe'_n}} + \sum \frac{i_n b_{\theta, \alpha} \theta t_n^{i_n} \cdots t_1^{i_1}}{1 - \theta t_n^{i_n} \cdots t_1^{i_1}} \neq 0$$

and there exist t'_n such that $t_n t_n'^{-1} \in F^{*p}$ and natural numbers r and $r_{\theta, \alpha}$ such that

$$\xi_{p^t} = (1 - \eta t_n'^{pe'_n})^r \prod_{\theta, \alpha} (1 - \theta t_n'^{i_n} t_{n-1}^{i_{n-1}} \cdots t_1^{i_1})^{r_{\theta, \alpha}},$$

where $r \equiv 0 \mod p^t$.

Thus,

$$\{\xi_p, t'_n, t_{n-1}, \ldots, t_1\} = p^{t-1}\{\xi_{p^t}, t'_n, \ldots, t_1\}$$
$$= p^{2t-1}\{1 - \eta t_n'^{pe'_n}, t'_n, t_{n-1}, \ldots, t_1\} = 0,$$

since $2t - 1 \geq s$ and the order of the element $p^s\{1 - \eta t_n'^{pe'_n}, t'_n, t_{n-1}, \ldots, t_1\}$ is prime to p. Finally,

$$\{\xi_p, t_n, t_{n-1}, \ldots, t_1\} = \{\xi_p, t'_n, t_{n-1}, \ldots, t_1\}$$
$$+ \{\xi_p, t_n \cdot t_n'^{-1}, t_{n-1}, \ldots, t_1\} = 0.$$

2. We proceed to the general case. The extension $k(\xi_{p^r})/k$ is totally ramified for any $r \geq 0$; hence, there exists a prime element π' in k such that $\omega_{p^t}(\{\xi_{p^t}, \pi'\}) = 1$, and so, as Merkur'ev showed in [3], $\{\xi_{p^t}, \pi'\} = 0$.

We take $\pi', t_{n-1}, \ldots, t_1$ as our system of parameters and make use of Lemma 7. We then find that

$$\{\xi_{p^t}, \alpha_2, \ldots, \alpha_{n+1}\} = \{\xi_{p^t}, t_{n-1}, \ldots, t_1, u\}, \quad \text{where } u \in (1 + \mathfrak{M}_F)^*.$$

We have $\omega_{p^t}(\{\xi_{p^t}, u\pi', t_{n-1}, \ldots, t_1\}) = 1$, and hence
$$\{\xi_p, u\pi', t_{n-1}, \ldots, t_1\} = 0,$$
from which we have $\{\xi_p, \alpha_2, \ldots, \alpha_{n+1}\} = 0$. The theorem is proved.

REMARK. Our results for multidimensional local fields of the third type basically characterize the K^*-groups of these fields. The fundamental papers [7], [8] of Kato contain information on the usual Milnor K-groups of n-dimensional local fields.

Bibliography

1. S. V. Vostokov, *On the theory of class fields of a multidimensional local field*, Dokl. Akad. Nauk SSSR **274** (1984), no. 4, 780–782; English transl. in Soviet Math. Dokl. **29** (1984).
2. V. G. Lomadze, *On the ramification theory for two-dimensional local fields*, Mat. Sb. **109** (1979), no. 3, 378–394; English transl. in Math. USSR-Sb. **37** (1980).
3. A. S. Merkur'ev, *The group K_2 for a local field*, Trudy Mat. Inst. Steklov. **165** (1984), 115–118; English transl. in Proc. Steklov Inst. Math. **165** (1985).
4. A. N. Parshin, *Abelian coverings of arithmetic schemes*, Dokl. Akad. Nauk SSSR **243** (1978), no. 4, 855–858; English transl. in Soviet Math. Dokl. **19** (1978).
5. _____, *Local class field theory*, Trudy Mat. Inst. Steklov. **165** (1984), 143–170; English transl. in Proc. Steklov Inst. Math. **165** (1985).
6. J. E. Carroll, *On the torsion in K_2 of local fields*, Lecture Notes in Math., vol. 342, Springer-Verlag, Berlin and New York, 1973, pp. 464–473.
7. K. A. Kato, *A generalization of local class field theory by using K-groups*. I, J. Fac. Sci. Univ. Tokyo Sect. IA Math. **26** (1979), no. 2, 303–376.
8. _____, *A generalization of local class field theory by using K-groups*. II, J. Fac. Sci. Univ. Tokyo Sect. IA Math. **27** (1980), no. 3, 603–683.

Translated by N. KOBLITZ

K-Theory of Demazure Models

A. A. KLYACHKO

Let us denote by $K(X) = K_*(X)$ the Grothendieck group of coherent algebraic sheaves on a variety X. We will consider the case when X is a complete nonsingular variety on which we have an action of a torus T and where X contains an open orbit isomorphic to T. We shall call such a variety the Demazure model of the torus T.

Let us recall some results about such models [4], [5]. Denote by $\widehat{T}$ the character module of the torus T and let N be the **Z**-module dual to $\widehat{T}$. For every affine open T-invariant subvariety $U \subset X$ we consider the sets

$$\widehat{T}(U) = \{\chi \in \widehat{T} | \chi \in \Gamma(U, O_U)\},$$
$$N(U) = \{n \in N | \forall \chi \in \widehat{T}(U), \, n(\chi) \geq 0\}.$$

The convex hull of the set $N(U)$ in $N \otimes \mathbf{R}$ is a polyhedral cone in $N \otimes \mathbf{R}$. If X is nonsingular, then every such cone is spanned by a part of a basis of the group N. The set of all polyhedral cones of the form $N(U)$ gives a simplicial decomposition of the space $N \otimes \mathbf{R}$. We shall call such a decomposition a fan associated with the model X and we shall write $\Sigma = \Sigma(X)$. A fan uniquely determines the Demazure model $X = X(\Sigma)$.

Every affine open T-invariant subvariety $U \subset X$ contains a unique orbit closed in U (the orbit of smallest dimension). This establishes a one-to-one correspondence between the simplices $\sigma \in \Sigma$ and the orbits O_σ of the torus T on X. In addition, the dimension of σ is equal to the codimension of the corresponding orbit O_σ. An orbit O_τ is contained in the closure of the orbit O_σ if and only if σ is the boundary of the simplex τ.

Let us denote by J_σ the sheaf of ideals defining the subvariety $\overline{O}_\sigma$ (the closure of the orbit O_σ).

1991 *Mathematics Subject Classification*. Primary 19E08.
Translation of Investigations in Number Theory. Arithmetic of Algebraic Varieties, Intercollegiate Scientific Collection, Izdat. Saratov Univ., Saratov, 1982, pp. 61–72.

PROPOSITION 1. *Let X be the Demazure model of the torus T. Then $K(X)$ as an abelian group is generated by the sheaves of ideals J_σ.*

PROOF. For every closed subscheme $Z \subset Y$ we have the exact sequence [1]
$$K(Z) \to K(Y) \to K(Y/Z) \to 0.$$

Let us denote by $\text{Sk}^d X$ the union of orbits of codimension d, and let $\overline{\text{Sk}^d X}$ be its closure in X. Then the sequence
$$K((\overline{\text{Sk}^{d+1} X})) \to K(\overline{\text{Sk}^d X}) \to K(\text{Sk}^d X) \to 0 \tag{1}$$
is exact. Since $K(T) \cong \mathbf{Z}$ for every torus T (and consequently for every orbit), $K(\text{Sk}^d X)$ is a free abelian group generated by the ideals of orbits of codimension d. From the exact sequence (1) by induction on d we obtain that $K(\overline{\text{Sk}^d X})$ is generated by ideals corresponding to orbits of codimension greater than or equal to d. In particular, $K(X) = K(\overline{\text{Sk}^0 X})$ is generated by ideals of orbits.

COROLLARY 1. *$K(X)$ is a finitely generated abelian group.*

COROLLARY 2. *$K(X)$ is generated by classes of complete T-sheaves of ideals.*

If the simplex $\sigma = \langle e_1, e_2, \ldots, e_d \rangle$ is spanned by the vectors $e_1, e_2, \ldots, e_d \in N$ then we can construct a supplementary complex of locally free sheaves (the Koszul complex):

$$0 \leftarrow O_X/J_\delta \xleftarrow{\varepsilon} O_X \xleftarrow{d} \bigoplus_i J_{e_i} \xleftarrow{d} \bigwedge^2 \left(\bigoplus_i J_{e_i}\right) \xleftarrow{d} \bigwedge^3 \left(\bigoplus_i J_{e_i}\right) \leftarrow \cdots,$$
$$d(s_1 \wedge s_2 \wedge \cdots \wedge s_k) = \sum_{i=1}^{k} (-1)^{i+1} s_i (s_1 \wedge \cdots \wedge \hat{s}_i \wedge \cdots \wedge s_k). \tag{2}$$

PROPOSITION 2. *The Koszul complex (2) is a locally free resolution of the structure sheaf O_X/J_σ of the closure of the orbit O_σ.*

PROOF. Indeed, the local representation of the orbit $\overline{O}_\sigma$ as an intersection of the orbits $\overline{O}_{e_i}$ of codimension one is equivalent to the representation of a vector subspace of codimension d as the intersection of d hyperplanes [5]. It is well known that in this case the Koszul complex is exact.

COROLLARY 1. *$K(X)$ is generated by invertible (in particular locally free) sheaves.*

Let $K^*(X)$ be the Grothendieck ring of the category of locally free coherent sheaves on X. Then the group $K(X) = K_*(X)$ is a $K^*(X)$-module, and the canonical homomorphism $K^*(X) \to K(X)$ is compatible with the module structure. Together with Corollary 1 this gives

COROLLARY 2. *The canonical map $K^*(X) \to K(X)$ is epimorphic and it defines on $K(X)$ the structure of a ring.*

Let us denote by $\langle \sigma, \tau \rangle$ the simplex in Σ (if there is one), the set of edges of which is the noninteracting union of the edges of the simplices σ and τ.

COROLLARY 3. *For the product of the classes of closures of orbits we have the relation*

$$\overline{O}_\sigma \cdot \overline{O}_\tau = \begin{cases} \overline{O}_{\langle \sigma, \tau \rangle} & \text{if } \langle \sigma, \tau \rangle \text{ is a simplex in } \Sigma, \\ 0 & \text{if } \sigma \text{ and } \tau \text{ are not faces of a simplex}. \end{cases} \quad (3)$$

In particular, if $\sigma = \langle e_1, e_2, \ldots, e_d \rangle$ is a simplex spanned by the vectors $e_i \in N$ then

$$\overline{O}_\sigma = \overline{O}_{e_1} \cdot \overline{O}_{e_2} \cdot \cdots \cdot \overline{O}_{e_d}. \quad (4)$$

PROOF. By a class of orbit $\overline{O}_\sigma$ in $K(X)$ we understand the class of its structure sheaf O_X/J_σ (that is in $K(X)$ by definition $\overline{O}_\sigma = 1 - J_\sigma$). From the Koszul complex (2) we obtain the following equalities in $K(X)$:

$$\overline{O}_\sigma = \sum_{k=0}^{d}(-1)^k \bigwedge^k \left(\bigoplus_i J_{e_i}\right) = \sum_{k=0}^{d}(-1)^k \sum_{i_1 < i_2 < \cdots < i_k} J_{e_{i_1}} J_{e_{i_2}} \cdots J_{e_{i_k}}$$

$$= \prod_{i=1}^{d}(1 - J_{e_i}) = \overline{O}_{e_1} \cdot \overline{O}_{e_2} \cdot \cdots \cdot \overline{O}_{e_d}.$$

This proves (4) and the top part of (3). The remaining part of (3) follows from the fact that if σ and τ are not faces of a single simplex then $\overline{O}_\sigma$ and $\overline{O}_\tau$ do not intersect.

The sheaves of ideals J_e that define orbits of codimension one in X (e is a simplex of dimension one in Σ) are the generators of the Picard group $\operatorname{Pic} X$. All the relations among these generators in $\operatorname{Pic} X$ are known [3]–[5] and they have the form

$$\prod_e J_e^{f(e)} = 1, \qquad f \in \operatorname{Hom}(N, \mathbf{Z}) = \widehat{T}. \quad (5)$$

The formulas (3), (4), and (5) allow us to compute the product of arbitrary orbits in $K(X)$. The computations are considerably simplified if we work in the associated graded ring with respect to the filtration defined by powers of the ideal $I = \operatorname{Ker}(K(X) \to K(T))$ (we identify the torus T with an open orbit in X). From the exact sequence (1) for $d = 0$ it follows that I is a subgroup generated by classes of orbits of codimension greater than or equal to one. Formula (4) shows that the ith power I^i is generated (as a group) by orbits of codimension greater than or equal to i.

If in formula (5) we substitute $J_e = (1 - \overline{O}_e)$, then we obtain a linear relation among orbits of codimension one in the associated graded ring:

$$\sum_e f(e)\overline{O}_e \equiv 0 \pmod{I^2}. \quad (6)$$

The following proposition describes the linear relations among classes of orbits of any dimension.

PROPOSITION 3. *Let $\sum a_\sigma \overline{O}_\sigma$ be a cycle of codimension k on $X = X(\Sigma)$. Then the following conditions are equivalent*:
a) $\sum a_\sigma \overline{O}_\sigma \equiv 0 \pmod{I^{k+1}}$;
b) *the cycle $\sum a_\sigma \overline{O}_\sigma$ is numerically equivalent to zero*;
c) *the cycle $\sum a_\sigma \overline{O}_\sigma$ can be represented as a sum of cycles of the form $\sum f_\tau(e) \overline{O}_{\langle \tau, e \rangle}$, where $f_\tau \in \operatorname{Hom}(N/\mathbf{Z}_\tau, \mathbf{Z})$ and the summation is extended over those one-dimensional simplices $e \in N$ for which $\langle \tau, e \rangle$ is a simplex from Σ.*

PROOF. The assertion b) → c) is evident. The implication c) → a) follows from relation (6) and from the equality

$$\sum_\tau f_\tau(e) \overline{O}_{\langle \tau, e \rangle} = \left(\sum_e f_\tau(e) \overline{O}_e \right) \cdot \overline{O}_\tau$$

(see Corollary 3 of Proposition 2). The proof of b) → c) is based on the following combinatorial lemma.

LEMMA. *In any simplicial decomposition of an n-dimensional cell there is an n-dimensional simplex such that among all of its faces not belonging to the boundary of the cell there is the smallest one.*

The fan Σ induces a simplicial decomposition of the $(n-1)$-dimensional sphere S^{n-1}. Let $E \subset S^{n-1}$ be a maximal $(n-1)$-dimensional cell on whose simplices the cycle $C = \sum a_\delta \cdot \sigma$ can be made equal to zero by adding summands of the form $\sum f_\tau(e) \langle \tau, e \rangle$. We shall consider that the cycle C itself already does not contain any simplices from E.

Let us consider the complementary cell E': $S^{n-1} = E \cup E'$, and the cells E and E' have a common boundary. If E' consists of a single simplex, then the cycle C is zero and there is nothing to prove. Otherwise, by our combinatorial lemma the cell E' contains an n-dimensional simplex of the form $\langle \delta, \tau \rangle$, where τ is the smallest internal simplex in $\langle \delta, \tau \rangle$. Any k-dimensional simplex from $\langle \delta, \tau \rangle$ entering into the cycle C contains τ, since in the opposite case it belongs to the boundary of the cell E' and consequently to the cell E. Therefore $\dim \tau \leq k$.

If $\dim \tau = k$, then τ is the unique simplex in $\langle \delta, \tau \rangle$ entering into the cycle C. Therefore the multiplicity of the simplex τ in C coincides with the intersection index $(\sum a_\sigma \overline{O}_\sigma \cdot \overline{O}_\delta)$ which is equal to zero by assumption. Therefore the cycle C is equal to zero on the simplices from $\langle \delta, \tau \rangle$, which contradicts the maximality of the cell E.

If however $\dim \tau < k$, then we consider all the $(k-1)$-dimensional faces τ' from $\langle \delta, \tau \rangle$ that contain τ. Since the simplex $\langle \delta, \tau \rangle$ is spanned by a basis of the group N, we may choose the function $f_{\tau'}$ to be equal to zero on τ' and such that adding the cycle $\sum_e f_{\tau'}(e) \langle \tau', e \rangle$ annihilates the cycle

C on all the k-simplices from $\langle \delta, \tau \rangle$ that contain τ. This does not change the cycle C on the remaining k-simplices from $\langle \delta, \tau \rangle$ and it remains zero on the cell E. Thus, also in this case, by adding summands of the form $\sum_e f_{\tau'}(e)\langle \tau', e \rangle$, the cycle C can be made equal to zero on the simplices from $\langle \delta, \tau \rangle$.

COROLLARY 1. $K_i(X) = I^i/I^{i+1}$ are free abelian groups of finite rank. In particular, $K(X)$ is a free abelian group.

Indeed, the group of classes of cycles modulo numerical equivalence has no torsion.

COROLLARY 2. *The intersection index induces an exact duality over* **Z** *between* $K_i(X)$ *and* $K_{n-i}(X)$.

Formally from the proposition proven it follows that the pairing between $K_i(X)$ and $K_{n-i}(X)$ is nondegenerate. To establish the duality over **Z** it is necessary to show that, if the indices of intersection of a cycle with rational coefficients with all the orbits are integers, then the cycle itself is equivalent to an integral cycle. To do this we have to consider the maximal cell $E \subset S^{n-1}$ on which the cycle can be made integral, and word by word repeat the considerations with the combinatorial lemma used in the proof of the proposition.

The main result of Proposition 3 can be written down as an exact sequence

$$A_{k,1} \xrightarrow{d} A_{k,0} \xrightarrow{\varepsilon} K_{n-k}(X) \to 0,$$

where $A_{k;0}$ is the group of k-dimensional chains $\sum_{\dim \sigma = k} a_\sigma \cdot \sigma$, $A_{k;1}$ is the set of formal sums of the form $\sum_{\dim \tau = k-1} f_\tau \tau$,

$$f_\tau \in \mathrm{Hom}(N/\mathbf{Z}\tau, \mathbf{Z}) \quad \text{and} \quad d\left(\sum f_\tau \tau\right) = \sum_e f_\tau(e)\langle \tau, e \rangle.$$

We extend this sequence to the resolvent group $K_{n-k}(X)$. We shall denote by $A_{k;i}$ the set of formal sums of the form $\sum_{\dim \sigma = k-i} f_\sigma(x_1, x_2, \ldots, x_i)\sigma$, where $f_\sigma(x_1, x_2, \ldots, x_i)$ is a multilinear skew-symmetric integral function on $N/\sigma\mathbf{Z}$. In particular, $A_{k;k} = \bigwedge^k \widehat{T}$. In general, $A_{k;i}$ is a free abelian group of rank $\binom{n-k+i}{i}|\mathrm{Sk}^{k-i}\Sigma|$, where $|\mathrm{Sk}^{k-1}\Sigma|$ is the number of simplices of dimension $(k-i)$ in the fan Σ.

Let us consider the complex

$$0 \to A_{k,k} \xrightarrow{d} A_{k,k-1} \xrightarrow{d} \cdots \to A_{k,1} \xrightarrow{d} A_{k,0} \xrightarrow{\varepsilon} K_{n-k}(X) \to 0, \qquad (7)$$

where

$$d(f_\sigma(x_1, x_2, \ldots, x_i)\sigma) = \sum_{\langle e, \sigma \rangle} f_\sigma(x_1, x_2, \ldots, x_{i-1}, e)\langle e, \sigma \rangle$$

(the summation is extended over all the one-dimensional simplices $e \in N$ for which $\langle e, \sigma \rangle$ is a simplex form Σ). From the skew-symmetry of the functions $f_\sigma(x_1, x_2, \ldots, x_i)$ it follows that $d^2 = 0$.

PROPOSITION 4. *The complex* (7) *is an acyclic resolvent of the group* $K_{n-k}(X)$.

PROOF. We again use the combinatorial lemma from the previous proposition.

We consider the element $\sum f_\sigma(x_1, x_2, \ldots, x_i)\sigma \in A_{k;i}$ belonging to the kernel of the differential d. Let $E \subset S^{n-1}$ be the maximal cell from which we can shift away the support of the sum $\sum f_\sigma(x_1, x_2, \ldots, x_i)\sigma$ by adding the differentials of elements of $A_{k;i+1}$; we will assume that already the initial sum $\sum f_\sigma \cdot \sigma$ does not contain simplices from E. We consider the complementary cell E', $S^{n-1} = E \cup E'$, $\dot{E} = \dot{E}' = E \cap E'$. According to the combinatorial lemma we can find an n-simplex in E of the form $\langle \delta, \tau \rangle$, where τ is the smallest interior simplex from $\langle \delta, \tau \rangle$. Since every simplex from $\langle \delta, \tau \rangle$ entering into the support of the sum $\sum f_\sigma(x_1, x_2, \ldots, x_i)\sigma$ does not lie on the face of the cell E, it must therefore contain τ. In particular, the differentials can contain simplices from $\langle \delta, \tau \rangle$ only if $\langle \delta, \tau_\sigma \rangle \supset \sigma \supset \tau$. Therefore from the condition $d(\sum f_\sigma(x_1, x_2, \ldots, x_i)\sigma) = 0$ we obtain that

$$\sum_{\sigma \supset \tau; \langle \sigma, e \rangle \supset \langle \delta, \tau \rangle} f_\sigma(x_1, x_2, \ldots, x_{i-1}, e)\langle \sigma, e \rangle = 0. \tag{8}$$

Let us connect with the simplex $\langle \delta, \tau \rangle$ a complex analogous to (7):

$$0 \to A_{k,m}(\langle \delta, \tau \rangle) \xrightarrow{d} A_{k,m-1}(\langle \delta, \tau \rangle) \to \cdots \xrightarrow{d} A_{k,1}(\langle \delta, \tau \rangle) \xrightarrow{d} A_{k,0}(\langle \delta, \tau \rangle), \tag{9}$$

where $m = k - \dim \tau$, $A_{k;i}(\langle \delta, \tau \rangle)$ is the set of formal sums of the form

$$\sum_{\tau \subset \sigma \subset \langle \delta, \tau \rangle} f_\sigma(x_1, \ldots, x_i)\sigma, \quad \dim \sigma = k - i,$$

$f_\sigma(x_1, \ldots, x_i)$ is a multilinear skew-symmetric function on $N/\mathbf{Z}\sigma$, and

$$d(f_\sigma(x_1, x_2, \ldots, x_i)\sigma) = \sum_{\langle \sigma, e \rangle \subset \langle \delta, \tau \rangle} f_\sigma(x_1, \ldots, x_{i-1}e)\langle \sigma, e \rangle.$$

The functions $f_\sigma(x_1, x_2, \ldots, x_i)$ are uniquely determined by their values $f_\sigma(e_1, e_2, \ldots, e_i)$, where $\langle e_1, \ldots, e_i \rangle$ is an oriented simplex from $\langle \delta, \tau \rangle$. This allows us to interpret $A_{k;i}(\langle \delta, \tau \rangle)$ as a group of oriented cochains of dimension $m - 1$ on the union of all the m-dimensional simplices from δ glued together at zero; $f_\sigma(x_1, \ldots, x_i)$ is then identified with the oriented cochain whose value on the boundary of the m-dimensional simplex $\langle \rho, \pi \rangle$ is equal to $f_\sigma(e_1, \ldots, e_i)$, where $\langle e_1, \ldots, e_i \rangle = \rho$.

One checks immediately that the differential that we introduced coincides with the differential of this simplicial complex and, consequently, the complex (9) is acyclic. Hence from formula (8) it follows that the initial chain $\sum f_\sigma \cdot \sigma$ can be shifted away from the simplex $\langle \delta, \tau \rangle$ by adding to it the differential of some chain $\sum f_{\sigma'} \cdot \sigma'$, $\tau \subset \sigma' \subset \langle \delta, \tau \rangle$. Since none of the simplices σ' belong to the boundary of the cell E', the resulting cycle remains zero on the complementary cell E. This contradicts the maximality of E and proves our proposition.

COROLLARY 1. *The rank of the group $K_m(X)$ (equal to the rank of the group $K_{n-m}(X)$) can be computed from the formula*

$$r(K_m(X)) = r(K_{n-m}(X)) = \sum_{i=0}^{n-m} (-1)^{n-m-i} \binom{n-i}{m} |\operatorname{Sk}^i \Sigma|.$$

Here $|\operatorname{Sk}^i \Sigma|$ is the number of simplices of the fan Σ of dimension i, and the summands $\binom{n-i}{m}|\operatorname{Sk}_i \Sigma|$ are the ranks of the groups $A_{n-m;n-m-i}$.

COROLLARY 2. *The rank of the group $K(X)$ is equal to $|\operatorname{Sk}^n \Sigma|$.*

PROOF.

$$r(K(X)) = \sum_{m=0}^{n} r(K_m(X)) = \sum_{m=0}^{n} \sum_{i=0}^{n-m} (-1)^{n-m-i} \binom{n-i}{m} |\operatorname{Sk}^i \Sigma|$$

$$= \sum_{i=0}^{n} \sum_{m=0}^{n-i} (-1)^{n-m-i} \binom{n-i}{m} |\operatorname{Sk}^i \Sigma| = |\operatorname{Sk}^n \Sigma|.$$

Until now we have considered the case when the torus T and its Demazure model $X(\Sigma)$ were defined over an algebraically closed field K. Let us now consider the forms of this model defined over the field k. In this case on the module $\widehat{T}$ of characters (and on the module dual to N) there acts the Galois group G, which transforms the fan Σ into itself. Conversely any representation of the Galois group in the module N compatible with the fan Σ defines some form of the model $X(\Sigma)$.

We will be interested in the action of the Galois group on the ring $K(X)$.

PROPOSITION 5 (the trace formula). *The values of a character of the representation of the Galois group in the ring $K(X)$ are given by*

$$\chi(g) = \sum_{\sigma^g = \sigma} \det\nolimits_{N/\mathbf{Z}\sigma}(1 - g), \tag{10}$$

where the summation is taken over all the g-invariant simplices from the fan Σ and the determinants of the transformation $(1 - g)$ are computed in the module $N/\mathbf{Z}\sigma$.

REMARK. It is enough to take the summation in the trace formula over the maximal g-invariant simplices. Indeed, if σ is contained in a larger g-invariant simplex $\langle \delta, \tau \rangle$, then τ will also be invariant. But then the sum of the edges of the simplex τ gives an invariant vector in $N/\mathbf{Z}\sigma$ and therefore $\det N/\mathbf{Z}\sigma(1 - g) = 0$.

PROOF OF THE PROPOSITION. Let us first of all notice that for every representation T of the group G the character of the virtual representation

$\sum(-1)^i \bigwedge^i T$ (the alternating sum of exterior powers) is equal to $\det_T(1-g)$. Let us consider the resolution (7):

$$0 \to A_{m;m} \to A_{m;m-1} \to \cdots \to A_{m;1} \to A_{m;0} \to K_{n-m}(X) \to 0.$$

In the Grothendieck ring of $\mathbf{Z}G$-modules it gives the relation

$$K_{n-m}(X) = \sum_0^m (-1)^j A_{m,j}. \tag{11}$$

Further, from the definition of the groups $A_{m;j}$ it follows that they can be written as a sum of induced modules

$$A_{m,j} = \sum_{\dim \sigma = (m-j)} \mathrm{Ind}_{G_\sigma}^G \bigwedge^j (N/\mathbf{Z}\sigma)^*, \tag{12}$$

where $G_\sigma^1 = \{g \in G | \sigma^g = \sigma\}$ is the stabilizer of the simplex σ, $\bigwedge^j(N/\mathbf{Z}\sigma)^*$ is the module of skew-symmetric multilinear functions on $N/\mathbf{Z}\sigma$, and the summation is taken over representatives of orbits of the group G on $\mathrm{Sk}^{m-j}\Sigma$.

From the relations (11) and (12) we obtain

$$\begin{aligned} K(X) &= \sum_0^n K_m(X) = \sum_{m=0}^n \sum_{j=0}^m (-1)^j A_{m,j} \\ &= \sum_{m=0}^n \sum_{j=0}^m (-1)^j \sum_{\dim \sigma = m-j} \mathrm{Ind}_{G_0}^G \bigwedge^j (N/\mathbf{Z}\sigma)^* \\ &= \sum_{\sigma \in \Sigma} \mathrm{Ind}_{G_\sigma}^G \left(\sum_j (-1)^j \bigwedge^j (N/\mathbf{Z}\sigma)^* \right). \end{aligned} \tag{13}$$

It remains only to notice that the value on the element g of the induced character $\mathrm{Ind}_{G_\sigma}^G (\sum_i (-1)^j (N/\mathbf{Z}\sigma)^*)$ is equal to the sum $\sum_{\rho^g = \rho} \det_{N/\mathbf{Z}\rho}(1-g)$ taken over all the g-invariant simplices ρ from the orbit of the simplex σ.

EXAMPLE. $\chi(-1) = 2^{\dim X}$.

COROLLARY 1. $\chi(g) > 0$ for all $g \in G$.

Indeed the eigenvalues of the transformation g are roots of unity and together with a root E an eigenvalue of the same multiplicity is also any conjugate over Q root ε'. Consequently the determinant $\det(1-g)$ is the product of the norms of elements of the form $1 - \varepsilon$. But

$$N(1-\varepsilon) = \begin{cases} p & \text{if } \varepsilon \text{ is a } p^\alpha\text{th primitive root of unity,} \\ 1 & \text{otherwise.} \end{cases}$$

Thus, $\det N/\mathbf{Z}\sigma(1-g) > 0$ for any maximal g-invariant simplex σ.

Let $K_Q(G)$ be the ring of rational characters of the group G and let $B(G) \subset K_Q(G)$ be the Burnside subring consisting of $\mathbf{Z}$-linear combinations of characters induced by trivial characters of subgroups.

COROLLARY 2. *If the torus T is isogenous to a rational torus (i.e., if the quotient space of the torus T by a finite group is birationally equivalent to a projective space) then the characters of the G-modules $K_m(X)$ belong to the Burnside ring $B(G)$.*

PROOF. It is known that if the torus T is isogenous to a rational torus then the character of the representation N (or $\widehat{T}$) lies in $B(G)$ [3]. But for every simplex σ the group $\mathbf{Z}\sigma$ is a transitive $G\sigma$-module ($G\sigma$ is the stabilizer of the simplex σ) and therefore the character $G\sigma$ of the module $N/\mathbf{Z}\sigma$ lies in the Burnside ring $B(G\sigma)$, but then the characters of the representations $(N/\mathbf{Z}\sigma)^*$ and $\bigwedge^j(N, \mathbf{Z}\sigma)^*$ are also in the ring $B(G\sigma)$. In its turn this implies that the characters of the induced representations $\mathrm{Ind}_{G_\sigma}^G(\bigwedge^j(N/\mathbf{Z}\sigma)^*)$ lie in $B(G)$. Finally, formulas (11) and (12) show that the characters of the representations $A_{m;j}$ and $K_{n-m}(X)$ lie in $B(G)$.

REMARKS. 1. The character of the representation G in the group $K(X)$ always belongs to the Burnside ring. Indeed, due to formula (13) it is enough to show that for every G-module M, the character of the virtual representation $\sum (-1)^j \bigwedge^j(M)$ belongs to the Burnside ring $B(G)$. But the exterior algebra of the module M is isomorphic to the homology of the topological torus $M \otimes \mathbf{R}/M$. Choose some G-invariant simplicial decomposition of the torus $M \otimes \mathbf{R}/M$. The corresponding groups of chains $C_i(M \otimes \mathbf{R}/M)$ are permutation modules; consequently, the character of the virtual representation

$$\sum (-1)^i C_i(M \otimes \mathbf{R}/M) = \sum (-1)^i H_i(M \otimes \mathbf{R}/M, \mathbf{Z}) = \sum (-1)^i \bigwedge^i (M)$$

belongs to the Burnside ring.

2. Analogously to how it was done for the Galois group, we can compute the trace of the Frobenius morphism in $K(X)$. If we denote by g the automorphism of the group N induced by the generator $\sigma: x \to x^q$ of the Galois group then

$$\mathrm{Tr}_{K(X)}(\mathrm{Fr}_q X) = \sum_{\sigma^g = \sigma\,;\, \sigma \in \Sigma} \det_{N/\mathbf{Z}\sigma}(q - g).$$

The terms of this sum coincide with the number of rational points on the orbits, and the whole sum is equal to the number of points in $X(\mathbf{F}_q)$. This leads to the idea that the odd-dimensional cohomology of X is equal to zero and that the homomorphisms $K_i(X) \to H^{2i}(X, \mathbf{Q}_p)$, induced by the characteristic classes, are isomorphisms.

BIBLIOGRAPHY

1. A. Borel and J.-P. Serre, *Le théorème de Riemann-Roch*, Bull. Soc. Math. France **86** (1958), 97–136.
2. Yu. I. Manin, *Lectures on the K-functor in algebraic geometry*, Uspekhi Mat. Nauk **24** (1969), no. 5, 3–86; English transl. in Russian Math. Surveys **24** (1969).

3. V. E. Voskresenskiĭ, *Algebraic tori*, "Nauka", Moscow, 1977. (Russian)
4. M. Demazure, *Sous-groupes algebriques de rang maximum du groupe de Cremona*, Ann. Sci. École Norm. Sup. (4) **3** (1970), 507–588.
5. G. Kempf et al., *Toroidal embeddings*. I, Lecture Notes in Math., vol. 339, Springer-Verlag, Berlin and New York, 1973.

Translated by P. BLASS

Norms in Fields of Real Algebraic Functions, and the Reduced Whitehead Group

UDC 513.6

V. A. LIPNITSKIĬ

Abstract. It is proved that norm mapping in finite-dimensional division algebras over the field k of the real algebraic functions satisfies Hasse's principle. This fact is used to calculate the Whitehead group for division algebras over fields of multiple series with coefficients from k.

Let R be the field of real numbers, and k the field of algebraic functions of one variable with coefficients in R. By $T(k)$ we denote the set of all real places of the field k, i.e., of places v mapping k on $R \cup \infty$, which are trivial on R; let k_v be the v-adic completion of k. If L is a ring, then L^* denotes the multiplicative group of this ring.

It is known that the Hasse principle ([6], Theorem 3.2.4) holds for finite Galois extensions F/k.

An element $\alpha \in k$ is the norm of an element of F if and only if $\alpha \in N_{k_v}^{F_w}(F_w^*)$ for all $v \in T(k)$ and all $w|v$, i.e., for all places w of F which extend v.

An analogous property has the reduced norm Nrd in finite-dimensional algebras with center k [7] (for a definition and properties of Nrd, see [1] or [2]).

Here the Hasse principle has, in fact, a more general significance.

THEOREM 1. *Let D be a Galois division algebra over k with center Z. An element $\alpha \in k$ belongs to the group $N_k^Z \operatorname{Nrd} D^*$ if and only if $\alpha \in N_{k_v}^{Z_w} \operatorname{Nrd} D_w^*$ for all $v \in T(k)$ and all places w of the field F which extend v.*

This result is used in the investigation of the reduced Whitehead group

---

1991 *Mathematics Subject Classification.* Primary 12E15; Secondary 19B14.
Translation of Dokl. Akad. Nauk BSSR **26** (1982), no. 7, 585–588.

$SK_1(A)$ of finite-dimensional simple algebras A. The role of this group in the theory of algebraic groups is well known (see, e.g., [3]–[5], where elements of the reduced K-theory of V. P. Platonov, devoted to the investigation of $SK_1(A)$, are also presented).

Let $K = k\langle u_1\rangle\langle u_2\rangle\cdots\langle u_m\rangle$ be the field of m-fold power series over k, and let A be a finite-dimensional simple algebra with center K. The following theorem holds.

THEOREM 2. *The reduced Whitehead group of the algebra A is trivial.*

It is sufficient to give a proof for division algebras A. Just as the field K, A is also a complete division algebra with respect to a discrete norm, and it has m successive residue algebras. We denote by D the mth residue division algebra. It is a Galois division algebra over k with the center Z abelian over k (Lemma 4 in [8]). Let $Z = L_0 \supset L_1 \supset \cdots \supset L_s = k$ be a decomposition of the extension Z/k into a tower of cyclic subextensions. We denote by $SL(1, A)$ the kernel of the reduced norm $\mathrm{Nrd}: A^* \to K^*$, and let A' be the commutator subgroup of A^*. Let $[B]_m$ be the mth reduction of elements $b \in B \subset A$, provided it is defined; $F_i = N_i([SL(1, A)]_m)$, where $N_i = N_{L_i}^{L_0}\mathrm{Nrd}_D$; $G_i = N_i([A']_m)$; P_j the subgroup of special projective conorms from $\Phi_j = L_{j-1}^*/L_j^* N_{j-1}(D^*)$, i.e., the subgroup of $a^* \in \Phi_j$ such that $a/\sigma(a) \in N_{j-1}(D^*)$ for $a \in a^*$, where σ is a generator of the Galois group $G(L_{j-1}/L_j)$.

The following theorem describes the structure of groups P_j.

THEOREM 3. *Under the assumptions of Theorem 1, let L be a subfield of Z cyclic over k. Then the subgroup P of special projective conorms from $\Phi = L^*/k^* N_L^Z \mathrm{Nrd}\, D^*$ is trivial.*

If D is not commutative, then, in view of the theorem of Tsen, it splits over the field $Z(\sqrt{-1})$, and consequently it has index 2 over Z. In both cases $SK_1(D) = 1$. If $SK_1(D)$ and all the groups P_j are trivial, then according to Corollary 3 of Theorem 3 of [8] we have $SK_1(A) \cong F_r/G_r$. But $F_r \subset k$ consists of roots of unity of degree $n \cdot n_0^{-1} \cdot [Z:k]^{-1}$, where n and n_0 are the indices of A and D, respectively, and $[Z:k]$ is the degree of the extension Z/k (Proposition 4, Chapter 1 of [8]). The field k contains only two roots of unity. Hence, either F_r is trivial (and then $SK_1(A) = 1$) or it has order 2. A simple argument shows that in the latter case F_r coincides with G_r, and, consequently, $SK_1(A)$ is also trivial.

In fact, if $F_r = \{\pm 1\}$, then the index n is even. $SK_1(A)$ is the direct product of the reduced Whitehead groups of primary components of the algebra A. Therefore one can assume that $n = 2^l$, where $l > 1$. Since D is either the algebra of quaternions over Z or it coincides with Z, one can find $0 \leq i < m$ such that the center $Z(A_i)$ of the $(m-i)$th reduction A_i of the algebra A is an extension of degree greater than 1 of the

$(m-i)$th reduction K_i of the field K. Since the extension $Z(A_i)/K_i$ is abelian (Lemma 4, Chapter 1 in [8]), it contains an intermediate field M_i of degree 2 over K_i. The field M_i can be lifted to a subfield $M \subset A$ of degree 2 over K. Hence $M = K(\sqrt{a})$ for some $a \in K$. In view of the Skolem-Noether theorem, the generator τ of the Galois group $G(M/K)$ can be extended to an inner automorphism φ_β of the algebra A. Then $-1 = \sqrt{a}/\tau(\sqrt{a'}) = \sqrt{a}, \beta(\sqrt{a})^{-1}, \beta^{-1} \in A'$, and consequently $-1 \in G_r$, i.e., $G_r = F_r$.

It remains to prove Theorems 1 and 3. Since in view of Tsen's theorem D splits over the field $Z(\sqrt{-1})$, it follows that D contains $Z(\sqrt{-1})$ as a maximal subfield. Taking into account that the characteristic of D is not 2, we can then choose in D a basis $1, i, j, ij$ such that $ji = -ij$, $i^2 = -1$, and $j^2 = \alpha$, where $\alpha \in Z^*/Z^{*2}$. Therefore, sometimes D is denoted by $(-1, \alpha)_Z$. The group $\operatorname{Nrd} D^*$ consists of elements which can be represented by the form $f = x_1^2 + x_2^2 - \alpha x_3^2 - \alpha x_4^2$. Then the algebra $(-1, \alpha)_Z$ is the full matrix algebra if and only if f represents zero. On the other hand, the triviality of D is equivalent to the triviality of D_v for all $v \in T(Z)$ [12]. This means that if D is not commutative, then the set $T(Z)$ (and consequently $T(k)$) is not empty. The topology of the field of real numbers induces a topology in $T(k)$, in which the mappings $x: v \to v(x) \in R \cup \infty$, putting in correspondence to every place $v \in T(k)$ the residue of x at the place v, are continuous for every $x \in k$. In this topology $T(k)$ is compact, and, in view of the theorem of Harnack, it consists of a finite number of connected components homeomorphic to a circle (see [9] or [11], Chapter VII, §4). These components are usually called ovals.

We shall need the following lemma.

LEMMA. *Let A and B be two finite-dimensional division algebras over a field k with centers Z and Z_1, respectively. Let $\varphi: A \to B$ be an isomorphism of algebras. If $N = N_k^Z \operatorname{Nrd}: A^* \to k^*$ and $N_1 = N_k^{Z_1} \operatorname{Nrd}: B^* \to k^*$, then $N_1(\varphi(\alpha)) = \varphi(N(\alpha))$ for every $\alpha \in A$. If φ is a k-isomorphism, then $N_1(\varphi(\alpha)) = N(\alpha)$, and consequently $N(A^*) = N_1(B^*)$.*

Let v be a place in $T(k)$. For any two extensions w_1 and w_2 of the place v onto Z there exists a σ in the Galois group $G(Z/k)$ such that $w_2 = \sigma w_1$ ([10], Chapter VI, Theorem 12). By continuity σ induces a k_v-isomorphism of fields $\tilde\sigma: Z_{w_1} \to Z_{w_2}$. On the other hand, since D is a Galois division algebra over k, σ can be extended to an automorphism θ of the division algebra D. Then $\varphi = \theta \otimes \tilde\sigma$ is an isomorphism of algebras $D_{w_1} = D \otimes_Z Z_{w_1}$ and $D_{w_2} = D \otimes_Z Z_{w_2}$. According to the Lemma the subgroups $N_{k_v}^{Z_{w_1}} \operatorname{Nrd} D_{w_1}^*$ and $N_{k_v}^{Z_{w_2}} \operatorname{Nrd} D_{w_2}^*$ of k_v^* coincide.

PROOF OF THEOREM 1. Let $\alpha \in N_k^Z \operatorname{Nrd} D^*$ and let $\beta \in D^*$ be such that $N_k^Z \operatorname{Nrd}(\beta) = \alpha$. We have $D = (-1, \varepsilon)_Z$ for some $\varepsilon \in Z^* \setminus Z^{*2}$.

The reduced norm in D_w can be described by the same quadratic form $f = x_1^2 + x_2^2 - \varepsilon x_3^2 - \varepsilon x_4^2$ as in the division algebra D. Consequently the element $\gamma = \mathrm{Nrd}_D(\beta)$ belongs to $\mathrm{Nrd}\, D_w^*$ for every $v \in T(k)$ and for all extensions w of the place v onto Z.

From the properties of the norm we have

$$\alpha = N_k^Z(\gamma) = \prod_{w|v} N_{k_v}^{Z_w}(\gamma) = \prod_{w|v} N_{k_v}^{Z_w} \mathrm{Nrd}_{D_w}(\beta) = \prod_{w|v} \alpha_w,$$

where $\alpha_w = N_{k_v}^{Z_w} \mathrm{Nrd}_{D_w}(\beta)$. As we have remarked above, all α_w belong to the same group $N_{k_v}^{Z_w} \mathrm{Nrd}\, D_w^* \subset k_v^*$. Consequently $\alpha \in N_{k_v}^{Z_w} \mathrm{Nrd}\, D_w^*$.

Now let α be a norm everywhere locally, i.e., $\alpha \in N_{k_v}^{Z_w} \mathrm{Nrd}\, D_w^*$ for every $v \in T(k)$ and for all extensions w of the place v onto Z. We shall find an element $\beta \in Z$ such that $\alpha^{-1} N_k^Z(\beta) \in k_v^{*2}$ for all $v \in T(k)$. Here we shall use the following fact which follows from Theorems 2 and 3 of [12]:

If on every oval in $T(k)$ there is given an even number of sign changes, then there exists a function $\delta \in k$ which attains the given signs at all places $v \in T(k)$.

Denote by H a finite set of places in $T(k)$ containing all zeros and poles of elements α belonging to the discriminant $D(Z/k)$, and such that $N_{Z/k}^{-1}(H)$ contains all zeros and poles of the element ε, and contains at least one place from every oval $O \subset T(Z)$. Under these conditions $T(Z) \backslash N_{Z/k}^{-1}(H)$ consists of intervals homeomorphic to the interval $(0, 1)$ of the real line, and the signs of the residues of ε are constant on every such interval.

Assume that for every place w in the interval $I' \subset T(Z) \backslash N_{Z/k}^{-1}(H)$ we have $w(\varepsilon) < 0$. Consequently, $\varepsilon = -1 \cdot d^2$ for some $d \in Z_w^*$, and $D_w = (-1, -1)_{Z_w}$ is the quaternion algebra. Then Nrd_w^*—the set of elements of Z_w^* represented by the form $g = x_1^2 + x_2^2 + x_3^2 + x_4^2$—coincides with Z_w^{*2}. Consequently

$$\alpha \in N_{k_v}^{Z_w} \mathrm{Nrd}\, D_w^* = N_{k_v}^{Z_w}(Z_w^{*2}) \subset k_v^{*2},$$

and, in particular, $v(\alpha) > 0$. In connection with this we attach the sign "+" to every interval I' such that D_w is a division algebra. We recall that for conjugate places w and σw, where $\sigma \in G(Z/k)$, the algebras D_w and $D_{\sigma w}$ are isomorphic. On the remaining intervals I' we define the signs in such a way that

$$\prod_{\sigma \in G(Z/k)} \mathrm{sign}\, \sigma(I') = \mathrm{sign}\, \alpha(v)/v \in N_{Z/k}(I').$$

Thus, on every oval there is given a finite number of sign changes. One can give only an even number of sign changes on a circle. Then there exists $\beta \in Z$ that takes the given signs on the intervals I'. According to Lemma

3.2.3 of [6]
$$\operatorname{sign} N_k^Z(\beta)(v) = \prod_{w|v} \operatorname{sign} \beta(w),$$

and, consequently, this sign coincides with $\operatorname{sign} \alpha(v)$. This means that $\delta = \alpha^{-1} N_k^Z(\beta)$ belongs to k_v^{*2} for all $v \in T(k)\setminus H$, i.e., $v(\delta) > 0$ or $v(\delta) = \infty$. The same property is had also by the residues of δ at the places $v \in H$, in view of the continuity of the mapping $\delta\colon v \to v(\delta)$. Such elements are called positively defined, and in view of Theorem 1 of [12] they have the form $a^2 + b^2$ for some $a, b \in k$. Evidently, the set of elements of this form coincides with the set $N_k^{k(i)}(k(i))$. It is easy to see that the norm mapping $N_k^Z \colon N_Z^{Z(i)}(Z(i)^*) \to N_k^{k(i)}(k(i)^*)$ is surjective. Therefore there exists an element $\gamma = c^2 + d^2 \in Z^*$, such that $N_k^Z(\gamma^{-1}) = \alpha^{-1} N_k^Z(\beta)$. Then $N_k^Z(\beta\gamma) = \alpha$. For every place $w \in T(Z)\setminus N_{Z/k}^{-1}(H)$, we have $\beta\gamma \in \operatorname{Nrd} D_w^*$. In fact, if D_w is a division algebra, then by construction $w(\beta\gamma) = w(\beta) > 0$, i.e., $\beta\gamma \in Z_w^{*2} \subset \operatorname{Nrd} D_w^*$, but in full matrix algebras the reduced norm coincides with the matrix determinant, and hence it is surjective. Then (see [7], Lemma 1.3 and Remark to it) there exists $\zeta \in D$ such that $\operatorname{Nrd}_D(\zeta) = \beta\gamma$, and, consequently, $N_k^Z \operatorname{Nrd}(\zeta) = \alpha$, which was to be proven.

REMARK. One can see from the proof that the sufficient condition of Theorem 1 can be weakened: one can disregard a finite number of places in $T(k)$.

The proof of Theorem 3 is based on Theorem 1 in basically the same way as were the proofs in the cases $D = Z$ or $D = L$ given in [7] and [8]; therefore it is omitted.

BIBLIOGRAPHY

1. N. Bourbaki, *Algèbre*, Chaps. 7–9, Actualités Sci. Indust., nos. 1179, 1261, 1272, Hermann, Paris, 1952, 1958, 1959.
2. Andre Weil, *Basic number theory*, Springer-Verlag, New York, 1967.
3. V. P. Platonov, *The Tannaka-Artin problem, and reduced K-theory*, Izv. Akad. Nauk SSSR Ser. Mat. **40** (1976), no. 2, 227–261; English transl. in Math. USSR-Izv. **10** (1976).
4. _____, *The Tannaka-Artin problem, and groups of projective conorms*, Dokl. Akad. Nauk SSSR **222** (1975), 1299–1302; English transl. in Soviet Math. Dokl. **16** (1975).
5. Jacques Tits, *Groupes de Whitehead de groupes algébriques simples sur un corps (d'apres V. P. Platonov et al.)*, Sèminaire Bourbaki, 29e annee (1976/77), Exp. No. 505, Lecture Notes in Math., vol. 677, Springer-Verlag, Berlin and New York, 1977, pp. 218–236.
6. J. T. Knight, *Riemann surfaces of field extensions*, Proc. Cambridge Philos. Soc. **65** (1969), 635–650.
7. V. A. Lipnitskii, *Triviality of the reduced Whitehead group over certain fields*, Mat. Zametki **24** (1978), no. 5, 629–640; English transl. in Math. Notes **24** (1978).
8. _____, *Reduced Whitehead group over special fields*, Summary of Candidate's Thesis, Minsk, 1980. (Russian)
9. S. Lang, *Some applications of the local uniformization theorem*, Amer. J. Math. **76** (1954), no. 2, 362–373.
10. Oscar Zariski and Pierre Samuel, *Commutative algebra*, vol. 2, Van Nostrand, Princeton, NJ, 1960.

11. I. R. Shafarevich, *Basic algebraic geometry*, "Nauka", Moscow, 1972; English transl., Springer-Verlag, Berlin and New York, 1974.
12. E. Witt, *Zerlegung reeller algebraischer Funktionen in Quadrate. Schiefkörper über reellem Funktionenkör*, J. Reine Angew. Math. **171** (1934), no. 1, 4–11.

V. I. Lenin Byelorussian State University

Received 8/DEC/81

Translated by J. BROWKIN

The Reduced Whitehead Group of Weakly and Totally Ramified Division Algebras

UDC 513.6; 519.4

V. A. LIPNITSKIĬ

> Abstract. An order of the reduced Whitehead group for weakly completely ramified division algebras over multiple Henzel discretely valued fields is exactly determined.

In recent years the reduced Whitehead group $SK_1(A)$ of finite-dimensional simple algebras A has been an object of active investigation (see, for example, [1]–[3]). The basic results, methods, and applications here are due to V. P. Platonov, Academician of the Academy of Sciences of the Belorussian SSR. In [3] he began an investigation of $SK_1(A)$ for algebras A over n-fold Henselian discretely valued fields. In particular, he determined that $SK_1(A)$ is finite and cyclic in the case of a commutative residue division algebra that is cyclic over the corresponding reduction of the center. In the present article we compute precisely the order of this group and establish a criterion for triviality of $SK_1(A)$ for the division algebras in the title.

Let K_m be an m-fold Henselian discretely valued field. Hence, m successive residue fields are defined: $K_{m-1} = \overline{K}_m$, $K_{m-2} = \overline{K}_{m-1}, \ldots, K_0 = \overline{K}_1$. Let $A = A_m$ be a finite-dimensional division algebra over K_m of index n, and let $A_{m-1} = \overline{A}_m, \ldots, A_0 = \overline{A}_1$ be the corresponding residue division algebras. We follow [3] in notation and terminology. It is assumed below that A is weakly and totally ramified over K_m, i.e., that $A_0 = K_0$ and $(n, \operatorname{char} K_0) = 1$.

Suppose that for an $a \in A$ the i-fold reduction ($1 \leq i \leq m$) is defined; denote it by $[a]_i$. For $B \subset A$ denote by $[B]_i$ the set of all $[a]_i$, $a \in B$, and let $[B]_i^1 = (a \in B | [a]_i = 1)$. A characteristic property of the set $[A]_i^1$ is the fact that the sth root can be taken for each element for every s not

---

1991 *Mathematics Subject Classification.* Primary 19B99; Secondary 16A39.
Translation of Vestsi Akad. Navuk BSSR Ser. Fiz.-Mat. Navuk **1986**, no. 2, 28–30.

©1992 American Mathematical Society
0065-9290/92 $1.00 + $.25 per page

divisible by the characteristic of K_0, and this sth root also belongs to $[A]_i^1$ (see [3], Lemma 2). Let A' be the commutant of the multiplicative group A^* of A, and let $SL(1, A)$ be the kernel of the reduced norm Nrd: $A^* \to K_m^*$. For every $a \in SL(1, A)$ the i-fold reduction of a is defined, and $[A]_i^1 \cap SL(1, A) \subset A'$ ([3], Theorem 1). Consequently, $SK_1(A)$ is isomorphic to the factor group $[SL(1, A)]_m/[A']_m$. In this case $[SL(1, A)]_m$ consists of all the nth roots of 1 contained in K_0. This is a cyclic group whose order d is a divisor of n in general. Let π_i be a prime element of the ring O_{A_i} of integers in A_i, $\varphi_{\pi_i}: x \to \pi_i x \pi_i^{-1}$ the corresponding inner automorphism of A_i, and σ_{ji} the $(i-j)$th reduction of φ_{π_i} acting on A_j, $0 < j < i$. A simple computation shows that the mth reduction of the commutant A' is generated by the elements

$$v_{ji} = [\pi_j/\sigma_{ji}(\pi_j)]_j \in [SL(1, A)]_m.$$

Among the subfields of the division algebra A are cyclic fields L over $F \supseteq K_m$ of the form $F(\sqrt[s]{b})$, $b \in F$. Let s be the largest of the degrees $[L : F]$ for such extensions. Obviously, $s \leq d$. We have the

THEOREM. *The order of the group $SK_1(A)$ is equal to the number $d \cdot s^{-1}$. In particular, $SK_1(A)$ is trivial for $s = d$.*

COROLLARY. *Suppose that K_0 contains all the nth roots of 1. Then $SK_1(A)$ is trivial if and only if A is a cyclic algebra.*

The proof of the theorem uses

PROPOSITION 1. *Suppose that L is a weakly and totally ramified extension of K_m. Then for every $\alpha \in L$ there exist $\beta \in [L]_m^1$ and a divisor r of the number $[L : K_m]$ such that $(\alpha\beta)^r \in K_m$.*

The proof of the proposition is by induction on m. We use known properties of Henselian fields (see [4]). The degree $[L : K_m]$ of the extension is $e \cdot f$, where e is the ramification index and $f = [\bar{l} : \overline{K}_m]$. The field L contains an unramified subextension M/K_m of degree f, and L is totally ramified over M. Let $\tilde{\pi}(u_m)$ be a prime element of the ring O_L (O_{K_m}), v_K a discrete valuation of the field K_m, and v_M (v_L) the extension of it to M (L). Since the extension M/K_m is unramified, the element u_m remains prime also in the ring O_M. As is known, $v_L(\tilde{\pi}^e) = v_L(u_m)$. Hence, $z = \tilde{\pi}^e \cdot u_m^{-1} \in O_L^*$. Then $\bar{z} \in \bar{L} = \overline{M}$, and $z \cdot \gamma^{-1} = \delta \in [L]_1^1$ for $\gamma \in M$ such that $\bar{\gamma} = \bar{z}$. As already mentioned, $\delta = \varepsilon^e$ for some $\varepsilon \in [L]_1^1$. Then $(\tilde{\pi} \cdot \varepsilon^{-1})^e = u_m \cdot \gamma \in M$, and $\pi = \tilde{\pi} \cdot \varepsilon^{-1}$ remains a prime element of O_L. Every element $\alpha \in L^*$ has the form $\pi^s w$ for some integer s, and $w \in O_L^*$. Since $\bar{w} \in \bar{L} = \overline{M}$, w can be represented as a product $w_1 \cdot w_2$, where $w_1 \in [L]_1^1$ and $w_2 \in M \cap O_L^* = O_M^*$. In this case

$$(\alpha w_1^{-1})^e = (\pi^s w_1 w_2 w_1^{-1})^e = (\pi^e)^s \cdot w_2^e = u_m^s \cdot \gamma^s \cdot w_2^e \in M$$

and the assertion is proved for $m = 1$. For $m > 1$ the induction hypothesis tells us that for the element $\overline{g} = \overline{\gamma^s \cdot w_2^e} \in \overline{M}$ there exist a $\overline{y} \in [\overline{M}]_{m-1}^1$ and a positive integer k dividing f such that $(\overline{z} \cdot y)^k \in \overline{K}_m$. Then $y \in [M]_m^1$ and $(zy)^k = \beta_1^k \cdot x$ for some $\beta_1 \in [M]_1^1$ and $x \in K_m$. Consequently, $(zy\beta_1^{-1})^k = x$. Since $y\beta_1^{-1} \in [M]_m^1$, $y \cdot \beta_1^{-1}$ has an eth root $\eta \in [M]_m^1$. In this case
$$(\alpha \cdot w_1^{-1} \cdot \eta)^{ek} = (u_m^s \cdot zy \cdot \beta_1^{-1})^k = u_m^{sk} \cdot x \in K_m.$$
The proposition is proved.

COROLLARY. *If A is a weakly and totally ramified division algebra of index n over the field K_m, then there exist prime elements π_i of the rings O_{A_i} that are roots of the binomial polynomials $f\pi_i(x) = x^{d_i} - a_i \in K_i[x]$, where $d_i | n$.*

The proof is obtained by applying the proposition to the extension $L_i = K_i(\pi_i)$ of the field K_i.

PROPOSITION 2. *Suppose that A is a weakly and totally ramified division algebra over the field K_m, $G_i = [A_i']_i$, $1 \leq i \leq m$, and G_m is a group of order μ. Then A contains the cyclic extension $L = F(\sqrt[\mu]{b})$ of degree μ of the field $F \supseteq K_m$, $b \in F$.*

It suffices to prove this for division algebras of primary index $n = p^l$, where p is a prime number. Assume that G_{r-1} is a proper subgroup of G_r for some r with $1 \leq r \leq m$. Let $v_{jr} = [\pi_j/\sigma_{jr}(\pi_j)]_j$, $1 \leq j < r$, a generator of G_r. This can be done because G_r is cyclic and its degree is primary. Assume also that π_j satisfies the corollary to Proposition 1. By assumption, the polynomial $x^\mu - 1$ factors into linear factors in K_0. The Hensel property gives us the same property for K_j. Hence, the inverse image ξ of the element v_{jr} in the field K_j can also be regarded as a μth root of 1. Then $\sigma_{jr}(\pi_j) = \xi\pi_j$, and hence $K_j(\pi_j)$ is invariant with respect to σ_{jr}. Let M be the subfield of $K_j(\pi_j)$ consisting of invariants of the group G of automorphisms generated by σ_{jr}. According to a basic theorem in Galois theory, the extension $K_j(\pi_j)/M$ is a cyclic extension with Galois group G, and the degree of this extension is equal to the order of G, i.e., the number μ. It is clear that this extension can be lifted to the required extension $L \supset F \supseteq K_m$ contained in A.

PROOF OF THE THEOREM. Suppose that the index of A is equal to n, d is the order of the group of nth roots of 1 contained in K_0, and A contains the cyclic extension $L = F(\sqrt[s]{b})$ of degree s of the field $F \supseteq K_m$. Then, by the Skolem-Noether theorem, the generator σ of the Galois group $G(L/F)$ extends to an inner automorphism φ_β of the algebra A. We have that $\sigma(\sqrt[s]{b}) = \xi\sqrt[s]{b}$ for some primitive sth root ξ of 1. Then $\xi = \sigma(\sqrt[s]{b})/\sqrt[s]{b} = \beta \cdot \sqrt[s]{b} \cdot \beta^{-1} \cdot (\sqrt[s]{b})^{-1} \in A'$, and $[\xi]_m$ is obviously a primitive sth root of 1. On the other hand, by Proposition 2, for every primitive μth root of 1 contained in $[A']_m$ there exists a cyclic extension $L = F(\sqrt[u]{b})$, $b \in F$, of the field $F \supseteq K_m$ that is contained in A. The assertion of the theorem can be obtained from this.

Bibliography

1. V. P. Platonov, *Algebraic groups and reduced K-theory*, Proc. Internat. Congr. Math. (Helsinki, 1978), Acad. Sci. Fennica, Helsinki, 1980, pp. 311–317.
2. P. Draxl and M. Kneser, SK_1 *von Schiefkörpern*, Lecture Notes in Math., vol. 778, Springer-Verlag, Berlin and New York, 1980.
3. V. P. Platonov, *On reduced K-theory for n-fold Hensel fields*, Dokl. Akad. Nauk SSSR **249** (1979), 1318–1320; English transl. in Soviet Math. Dokl. **20** (1979).
4. O. F. Schilling, *The theory of valuations*, Amer. Math. Soc., Providence, RI, 1950.

Belorussian State University

Received 4/APR/83

Translated by H. H. McFADEN

The Structure of Unitary and Orthogonal Groups Over Henselian Discretely Normed Azumaya Algebras

UDC 513.6

V. I. KASKEVICH

> Abstract. Unitary and reduced unitary Whitehead groups of Azumaya algebras over Henselian discrete valuation rings are calculated.

Let A be a finite-dimensional simple algebra over its center K. The negative solution of the Tannaka-Artin problem concerning the triviality of the reduced Whitehead groups $SK_1(A)$, given by V. P. Platonov [1], [2], led to the need to compute them. For that purpose reduced K-theory was created [3], [4], followed by its unitary analog [5]–[7], for computation of the corresponding reduced unitary Whitehead groups $SK_1U(\tau, A)$, where τ is an involution on A. The introduction of reduced K-theory and reduced unitary K-theory, apart from various applications, made possible the computation of those groups for a wider class of rings A. In [8] V. I. Yanchevskiĭ and the author calculated the functors K_1 and SK_1 for Azumaya algebras over Henselian discretely normed rings. The goal of the present paper is the calculation of the corresponding unitary functor K_1U and SK_1U for the above algebras.

Let K be a Henselian discretely normed field with ring of integer elements $\mathfrak{D}_K$, and let A be an Azumaya algebra over $\mathfrak{D}_K$. Then $A \cong M_r(\mathfrak{D}_D)$ for some natural number r and for the ring $\mathfrak{D}_D$ of the integer elements of a skew field D with center K. We shall suppose that an involution τ acts on the algebra A, induced by an involution τ of the ring $\mathfrak{D}_D$ such that for elements (a_{kl}) of A, $a_{kl} \in \mathfrak{D}_D$, $(a_{kl})^\tau = (a_{lk}^\tau)$. Similarly, an involution of A induces an involution on the matrix algebra $M_n(A)$, such that its action on $M_n(A) = M_{nr}(\mathfrak{D}_D)$ can be considered as an action of the involution induced in the

---

1991 *Mathematics Subject Classification*. Primary 19B28; Secondary 20G35.
Translation of Dokl. Akad. Nauk **31** (1987), no. 12, 1073–1076.

above manner from the ring $\mathfrak{D}_D$. The definitions and notations concerning unitary and orthogonal groups over a ring will be those in [9]. Let ε_r be an element in the center of the algebra A, such that $\varepsilon_r \cdot \varepsilon_r^\tau = 1_r$ is the identity element of A. Then $\varepsilon_r = \varepsilon \cdot 1_r$, where ε is an element in the center of the ring $\mathfrak{D}_D$ such that $\varepsilon \cdot \varepsilon^\tau = 1$ is the identity element of the ring $\mathfrak{D}_D$.

We denote by $E_{k,l}$ the matrix in $M_n(A)$, which has the (k, l)th entry equal to 1_r and all other entries equal to zero. Similarly, we denote by $e_{i,j}$ the matrix from $M_{nr}(\mathfrak{D}_D)$ which has its (i, j)th entry equal to 1 and the other entries equal to zero. We define

$$\varepsilon_{rn} = \sum_{k=1}^{n} \varepsilon_r E_{k,k} = \sum_{i=1}^{rn} \varepsilon e_{i,i}, \quad Q_n = \sum_{k=1}^{n} E_{2k-1,2k} \in M_{2n}(A),$$

$$F_n^{\varepsilon_r} = Q_n + \varepsilon_{2nr} Q_n^\tau \in GL(2n, A) = GL(2nr, \mathfrak{D}_D),$$

$$U^{\varepsilon_r}(2n, \tau, A) = \{X \in GL(2n, A) | X^\tau F_n^{\varepsilon_r} X = F_n^{\varepsilon_r}\},$$

$$SU^{\varepsilon_r}(2n, \tau, A) = U^{\varepsilon_r}(2n, \tau, A) \cap SL(2n, A),$$

$$M_{2n}(A)_\varepsilon = \{X - \varepsilon_{2nr} X^\tau | X \in M_{2n}(A)\},$$

$$O^{\varepsilon_r}(2n, \tau, A) = \{X \in GL(2n, A) | X^\tau Q_n X - Q_n \in M_{2n}(A)_\varepsilon\},$$

$$SO^{\varepsilon_r}(2n, \tau, A) = O^{\varepsilon_r}(2n, \tau, A) \cap SL(2n, A).$$

The main results (beginning with Proposition 1) are obtained under the assumption that $\operatorname{char} \overline{K} \neq 2$. In this case (see [9]) the orthogonal groups $O^{\varepsilon_r}(2n, \tau, A)$ and $SO^{\varepsilon_r}(2n, \tau, A)$ coincide with the corresponding unitary groups and for that reason we consider only the latter ones. We note additionally that if the involution τ is the identity, then $A = \mathfrak{D}_K$ and $U^{-1_r}(2n, \tau, A) = Sp(2n, A)$ is the standard symplectic group.

If we define similarly the unitary groups over $\mathfrak{D}_D$, then the following isomorphisms hold.

THEOREM 1.

$$U^{\varepsilon_r}(2n, \tau, A) \cong U^{\varepsilon}(2nr, \tau, \mathfrak{D}_D),$$
$$SU^{\varepsilon_r}(2n, \tau, A) \cong SU^{\varepsilon}(2nr, \tau, \mathfrak{D}_D).$$

This allows one to study the unitary groups over $\mathfrak{D}_D$.
According to [9] we have

THEOREM 2. *If* $n \geq 2$
(i)
$$U^{\varepsilon}(2n, \tau, \mathfrak{D}_D) = E^{\varepsilon}(2n, \tau, \mathfrak{D}_D) U^{\varepsilon}(2n-2, \tau, \mathfrak{D}_D),$$
where $E^{\varepsilon}(2n, \tau, \mathfrak{D}_D)$ *is the group generated by the unitary transvections;*
(ii)
$$K_1 U^{\varepsilon}(\tau, \mathfrak{D}_D) = U^{\varepsilon}(2n, \tau, \mathfrak{D}_D) / E^{\varepsilon}(2n, \tau, \mathfrak{D}_D)$$
$$= U^{\varepsilon}(2, \tau, \mathfrak{D}_D) / E^{\varepsilon}(2, \tau, \mathfrak{D}_D) [U^{\varepsilon}(2, \tau, \mathfrak{D}_D), U^{\varepsilon}(2, \tau, \mathfrak{D}_D)].$$

Denote by $T_\varepsilon(\tau, \mathfrak{D}_D)$ the group generated by the elements $1 + x^\tau y \in \mathfrak{D}_D^* = GL(1, \mathfrak{D}_D)$, where $x = -\varepsilon x^\tau \in \mathfrak{D}_D$ and $y = -\varepsilon y^\tau \in \mathfrak{D}_D$, and by $T(\tau, \mathfrak{D}_D)$ the group generated by the elements zz^τ, where $z \in \mathfrak{D}_D^*$. Then

(iii) *if there exists* $y = -\varepsilon y^\tau \in \mathfrak{D}_D^*$, *then*

$$K_1 U^\varepsilon(\tau, \mathfrak{D}_D) = \mathfrak{D}_D^*/T_\varepsilon(\tau, \mathfrak{D}_D) T(\tau, \mathfrak{D}_D) [\mathfrak{D}_D^*, \mathfrak{D}_D^*];$$

(iv) *if there is no* $y = -\varepsilon y^\tau \in \mathfrak{D}_D^*$, *then* $\varepsilon - 1 \in \mathfrak{P}_D$, *where* $\mathfrak{P}_D$ *is a maximal ideal of the ring* $\mathfrak{D}_D$, *the involution is the identity on* $\overline{D} = \mathfrak{D}_D/\mathfrak{P}_D$, *and*

$$K_1 U^\varepsilon(\tau, \mathfrak{D}_D) = \mathfrak{D}_D^*/T_\varepsilon(\tau, \mathfrak{D}_D) T(\tau, \mathfrak{D}_D) [\mathfrak{D}_D^*, \mathfrak{D}_D^*] \times \mathbf{Z}/2\mathbf{Z}.$$

Analogous assertions are valid also for the special unitary groups:

THEOREM 3. *If* $n \geq 2$
(i) $SU^\varepsilon(2n, \tau, \mathfrak{D}_D) = E^\varepsilon(2n, \tau, \mathfrak{D}_D) SU^\varepsilon(2n - 2, \tau, \mathfrak{D}_D)$;
(ii)

$$SK_1 U^\varepsilon(\tau, \mathfrak{D}_D) = SU^\varepsilon(2n, \tau, \mathfrak{D}_D)/E^\varepsilon(2n, \tau, \mathfrak{D}_D)$$
$$= SU^\varepsilon(2, \tau, \mathfrak{D}_D)/E^\varepsilon(2, \tau, \mathfrak{D}_D) [U^\varepsilon(2, \tau, \mathfrak{D}_D), U^\varepsilon(2, \tau, \mathfrak{D}_D)];$$

(iii) *if there exists* $y = -\varepsilon y^\tau \in \mathfrak{D}_D^*$ *or if there is no* $z \in \mathfrak{D}_D^*$ *such that* $\mathrm{Nrd}_D(z) = \mathrm{Nrd}_D(-\varepsilon z^\tau)$, *then*

$$SK_1 U^\varepsilon(\tau, \mathfrak{D}_D) = \Sigma'_\tau(\mathfrak{D}_D)/T_\varepsilon(\tau, \mathfrak{D}_D) T(\tau, \mathfrak{D}_D) [\mathfrak{D}_D^*, \mathfrak{D}_D^*],$$

where $\Sigma'_\tau(\mathfrak{D}_D)$ *is the group generated by those elements* $x \in \mathfrak{D}_D^*$ *for which* $\mathrm{Nrd}_D(x) = \mathrm{Nrd}_D(x^\tau)$;

(iv) *if there exists* $z \in \mathfrak{D}_D^*$ *such that* $\mathrm{Nrd}_D(z) = \mathrm{Nrd}_D(-\varepsilon z^\tau)$ *and there is no* $y = -\varepsilon y^\tau \in \mathfrak{D}_D^*$, *then*

$$SK_1 U^\varepsilon(\tau, \mathfrak{D}_D) = \Sigma'_\tau(\mathfrak{D}_D)/T_\varepsilon(\tau, \mathfrak{D}_D) T(\tau, \mathfrak{D}_D) [\mathfrak{D}_D^*, \mathfrak{D}_D^*] \times \mathbf{Z}/2\mathbf{Z}.$$

For convenience in formulating the following statements we shall omit the direct factor $\mathbf{Z}/2\mathbf{Z}$ in the conditions of (iv).

COROLLARY 1. *Let* τ *and* μ *be involutions such that* $\mu = \tau i_u$, *where* i_u *is an inner automorphism of the skew field* D, *generated by an element* $u \in \mathfrak{D}_D^*$. *If* $u = u^\tau$, *then* $K_1 U^\varepsilon(\mu, \mathfrak{D}_D) = K_1 U^\varepsilon(\tau, \mathfrak{D}_D)$ *and* $SK_1 U^\varepsilon(\mu, \mathfrak{D}_D) = SK_1 U^\varepsilon(\tau, \mathfrak{D}_D)$. *If* $u = -u^\tau$, *then* $K_1 U^\varepsilon(\mu, \mathfrak{D}_D) = K_1 U^{-\varepsilon}(\tau, \mathfrak{D}_D)$ *and* $SK_1 U^\varepsilon(\mu, \mathfrak{D}_D) = SK_1 U^{-\varepsilon}(\tau, \mathfrak{D}_D)$.

Let us suppose now that the center $Z(\overline{D})$ of the skew field $\overline{D}$ is separable over $\overline{K} = \mathfrak{D}_K/\mathfrak{P}_K$, where $\mathfrak{P}_K$ is a maximal ideal in the ring $\mathfrak{D}_K$.

PROPOSITION 1. $(1 + \mathfrak{P}_D) \cap \Sigma'_\tau(\mathfrak{D}_D) \subset T(\tau, \mathfrak{D}_D)[\mathfrak{D}_D^*, \mathfrak{D}_D^*]$.

The proof is based on the application of a congruence-theorem from [8]. Proposition 1 is essential and allows us, by use of Theorems 2 and 3, to reduce the study of the groups $K_1 U^\varepsilon(\tau, \mathfrak{D}_D)$ and $SK_1 U^\varepsilon(\tau, \mathfrak{D}_D)$ to that of the corresponding groups for the skew field of residues $\overline{D}$ and of some groups (which in a special case are contained in $Z(\overline{D})^*$) which are deviations from the latter. Since the subgroups $T_\varepsilon(\tau, \mathfrak{D}_D)$ and $[\mathfrak{D}_D^*, \mathfrak{D}_D^*]$ are normal in $\mathfrak{D}_D^*$ we have

$$\overline{T_\varepsilon(\tau, \mathfrak{D}_D)T(\tau, \mathfrak{D}_D)[\mathfrak{D}_D^*, \mathfrak{D}_D^*]} = \overline{T_\varepsilon(\tau, \mathfrak{D}_D)}\,\overline{T(\tau, \mathfrak{D}_D)}\,\overline{[\mathfrak{D}_D^*, \mathfrak{D}_D^*]}.$$

Moreover, it is easy to see that $\overline{T(\tau, \mathfrak{D}_D)} = T(\overline{\tau}, \overline{D})$, $\overline{[\mathfrak{D}_D^*, \mathfrak{D}_D^*]} = [\overline{D}^*, \overline{D}^*]$, and the next proposition follows.

PROPOSITION 2. $\overline{T_\varepsilon(\tau, \mathfrak{D}_D)} = T_{\overline{\varepsilon}}(\overline{\tau}, \overline{D})$.

The next two theorems are basic.

THEOREM 4. *There exists an exact sequence*

$$1 \to (1 + \mathfrak{P}_K)^{1-\tau} \to K_1 U^\varepsilon(\tau, \mathfrak{D}_D) \to K_1 U^{\overline{\varepsilon}}(\overline{\tau}, \overline{D}) \to 1.$$

THEOREM 5. *The group $SK_1 U^\varepsilon(\tau, \mathfrak{D}_D)$ can be included in an exact sequence*

$$1 \to SK_1 U^{\overline{\varepsilon}}(\overline{\tau}, \overline{D}) \to SK_1 U^\varepsilon(\tau, \mathfrak{D}_D) \to N \to 1,$$

where $N = (\mathrm{Nrd}_{\overline{D}}(\overline{D}^*))^{1-\overline{\tau}} \cap N^{(1)}_{Z(\overline{D})/\overline{K}}$, $N^{(1)}_{Z(\overline{D})/\overline{K}} = \{x \in Z(\overline{D}) | N_{Z(\overline{D})/\overline{K}}(x) = 1\}$.

COROLLARY 2. *If τ is an involution of the first kind, then $K_1 U^\varepsilon(\tau, \mathfrak{D}_D) \cong K_1 U^{\overline{\varepsilon}}(\overline{\tau}, \overline{D})$.*

COROLLARY 3. $SK_1 Sp(A) = K_1 Sp(A) = K_1 Sp(\overline{K}) = 1$.

COROLLARY 4. *If the skew field D is not ramified or $\overline{\tau}$ is an involution of the first kind, then $SK_1 U^\varepsilon(\tau, \mathfrak{D}_D) \cong SK_1 U^{\overline{\varepsilon}}(\overline{\tau}, \overline{D})$.*

Theorems 4 and 5 are complemented by

PROPOSITION 3. (i) *If $\overline{\varepsilon} \neq 1$, then*

$$K_1 U^{\overline{\varepsilon}}(\overline{\tau}, \overline{D}) = K_1 U^{-1}(\overline{\tau}, \overline{D}) = \overline{D}^*/\Sigma_{\overline{\tau}}(\overline{D})[\overline{D}^*, \overline{D}^*],$$
$$SK_1 U^{\overline{\varepsilon}}(\overline{\tau}, \overline{D}) = SK_1 U^{-1}(\overline{\tau}, \overline{D}) = \Sigma'_{\overline{\tau}}(\overline{D})/\Sigma_{\overline{\tau}}(\overline{D})[\overline{D}^*, \overline{D}^*],$$

where $\Sigma_{\overline{\tau}}(\overline{D})$ is the group generated by the elements $x \in \overline{D}^*$ with $x = x^\tau$;
(ii) *if $\overline{\varepsilon} = 1$ and $\overline{\tau} = \mathrm{id}_{\overline{D}}$, then the skew field $\overline{D}$ is commutative and $K_1 U^1(\overline{\tau}, \overline{D}) = SK_1 U^1(\overline{\tau}, \overline{D}) = \overline{D}^*/\overline{D}^{*2}$ is a group of exponent ≤ 2.*

REMARK. If $\overline{\varepsilon} = 1$ and $\overline{\tau} \neq \mathrm{id}_{\overline{D}}$, then there exists an element $u \in \mathfrak{D}_D^*$, such that $u = -u^\tau$. Therefore, going over (see Corollary 1) to the groups $K_1 U^{-\varepsilon}(\tau i_u, \mathfrak{D}_D)$ and $SK_1 U^{-\varepsilon}(\tau i_u, \mathfrak{D}_D)$, we find ourselves in conditions (i) of Proposition 3.

In view of the above results we shall suppose now that the skew field D is ramified, $\overline{\tau}$ is an involution of the second kind, and $\varepsilon = -1$. Moreover, we shall write $SK_1 U$ instead of $SK_1 U^{-1}$.

PROPOSITION 4. *Let τ be an involution of the second kind and let K_τ be its field of invariants. The extension K/K_τ is fully ramified if and only if the restriction $\bar\tau$ on $Z(\overline{D})$ coincides with σ—the generator of the group $\mathrm{Gal}(Z(\overline{D})/\overline{K})$. If one of the last two equivalent conditions is satisfied, and also in the case where τ is an involution of the first kind, in the notation of Theorem 5 we have $N = \mathrm{Nrd}_{\overline{D}}(\overline{D}^*)^{1-\sigma}$.*

Now let τ be an involution of the second kind and K/K_τ be not ramified. For the formulation and proofs of further results the following lemma is important.

LEMMA 1. *There exists a skew field of inertia $B \subset D$, a simple element $\pi \in \mathfrak{D}_D$, and an involution τ_1, which satisfy the fixed point lemma ([5], 3.17) and are such that $SK_1U(\tau, \mathfrak{D}_D) = SK_1U(\tau_2, \mathfrak{D}_D)$, where $\tau_2 = \tau_1 i_\pi$.*

A group of projective unitary conorms is defined as follows. An element $\tilde a \in Z(\overline{D})^*/\overline{K}^* \mathrm{Nrd}_{\overline{D}}(\Sigma_{\bar\tau_1}(\overline{D}))$ is called a projective unitary norm if there exists $a \in \tilde a$ and $b \in \mathrm{Nrd}_{\overline{D}}(\overline{D}^*)$, such that $a^{\sigma-1} = b^{1-\bar\tau_2}$ (σ is a generator of the group $\mathrm{Gal}(Z(\overline{D})/\overline{K})$ and is the restriction of $\bar i_\pi$ on $Z(\overline{D})$). The collection of projective unitary norms forms a group, denoted by $PU(\tau_1, D)$.

PROPOSITION 5. *If $PU(\tau_1, D) = 1$, then $N = \mathrm{Nrd}_{\overline{D}}(\Sigma_{\bar\tau_1}(\overline{D}))^{1-\sigma}$.*

COROLLARY 5. *If the skew field $\overline{D}$ is commutative, or if $\overline{K}$ is a C_2^0-field, then $SK_1U(\tau, \mathfrak{D}_D) = (Z(\overline{D})_{\bar\tau_1^*})^{1-\sigma}$.*

COROLLARY 6. *Let $\overline{K}$ be a global field or a field of real algebraic functions of one variable. Then*

$$SK_1U(\tau, \mathfrak{D}_D) = (Z(\overline{D})_{\bar\tau_1^*} \cap \mathrm{Nrd}_{\overline{D}}(\overline{D}^*))^{1-\sigma}.$$

The author is grateful to V. I. Yanchevskiĭ for useful discussions and advice.

BIBLIOGRAPHY

1. V. P. Platonov, *On the Tannaka-Artin problem*, Dokl. Akad. Nauk SSSR **221** (1975), 1038–1041; English transl. in Soviet Math. Dokl. **16** (1975).
2. _____, *The Tannaka-Artin problem, and groups of projective conorms*, Dokl. Akad. Nauk SSSR **222** (1975), 1299–1302; English transl. in Soviet Math. Dokl. **16** (1975).
3. _____, *The Tannaka-Artin problem, and reduced K-theory*, Izv. Akad. Nauk SSSR Ser. Mat. **40** (1976), no. 2, 227–261; English transl. in Math. USSR-Izv. **10** (1976).
4. _____, *Reduced K-theory and approximation in algebraic groups*, Trudy Mat. Inst. Steklov. **142** (1976), 198–207; English transl. in Proc. Steklov Inst. Math. **142** (1979).
5. V. I. Yanchevskiĭ, *Reduced unitary K-theory and division algebras over Henselian discretely valued fields*, Izv. Akad. Nauk SSSR Ser. Mat. **42** (1978), no. 4, 879–918; English transl. in Math. USSR-Izv. **13** (1979).
6. _____, *Reduced unitary K-theory. Applications to algebraic groups*, Mat. Sb. **110** (1979), no. 4, 579–596; English transl. in Math. USSR-Sb. **38** (1981).
7. V. P. Platonov and V. I. Yanchevskiĭ, *Dieudonne's conjecture on the structure of unitary groups over a skew-field and Hermitian K-theory*, Izv. Akad. Nauk SSSR Ser. Mat. **48** (1984), no. 6, 1266–1294; English transl. in Math. USSR-Izv. **25** (1985).

8. V. I. Kaskevich and V. I. Yanchevskiĭ, *The structure of the general linear group and the special linear group over Henselian discretely valued Azumaya algebras*, Dokl. Akad. Nauk BSSR **31** (1987), 5–8. (Russian)
9. L. N. Vaserstein, *Stabilization of unitary and orthogonal groups over a ring with involution*, Mat. Sb. **81** (1970), no. 3, 328–351; English transl. in Math. USSR-Sb. **10** (1970).

Institute of Mathematics
Academy of Sciences of the Belorussian SSR

Received 12/JAN/87

Translated by O. MACEDONSKA-NOSALSKA

Reduced Unitary Whitehead Groups and Noncommutative Rational Functions

UDC 513.6

V. I. YANCHEVSKIĬ

Abstract. The reduced unitary Whitehead groups of skew fields of noncommutative rational functions are calculated.

In connection with various needs of algebraic K-theory and the theory of algebraic groups, V. P. Platonov [1] developed the reduced K-theory which studies the reduced Whitehead groups $SK_1(A)$ of finite-dimensional division algebras A. Ideas and methods of this work allowed us to compute $SK_1(A)$ for division algebras whose centers belong to a wide class of fields including all henselian fields. In the case when the centers are fields of algebraic functions in several variables, until recently one could obtain only qualitative results on computation of the groups $SK_1(A)$. However in the works [2], [3], using new ideas, the groups $SK_1(A)$ were computed for so-called division rings of noncommutative rational functions in several variables.

In the present paper arguments similar to those of [2], [3] are used to obtain results on computation of the reduced unitary Whitehead groups (the main subject of study in the reduced unitary K-theory) for division rings of noncommutative rational functions with involutions.

We recall definitions we need. Let T be a finite-dimensional central division algebra over $Z(T)$ with an involution τ of the second kind. We set $S_\tau = \{t \in T | t^\tau = t\}$, $Z(T)_\tau = Z(T) \cap S_\tau$, and $T^* = T\backslash 0$. Let $\Sigma'_\tau(T) = \{t \in T^* | \mathrm{Nrd}_T(t) \in Z(T)_\tau\}$ (here Nrd_T is the reduced norm homomorphism), and let $\Sigma_\tau(T)$ be the subgroup of T^* generated by the elements of $S^*_\tau = S_\tau\backslash 0$. Then the factor group $\Sigma'_\tau(T)/\Sigma_\tau(T)$ is called the reduced unitary Whitehead group $SUK_1(\tau, T)$ (see [4] for details); $SL(1, T) = \{t \in T^* | \mathrm{Nrd}_T(t) = 1\}$.

1991 *Mathematics Subject Classification.* Primary 19B28; Secondary 20G35.
Translation of Dokl. Akad. Nauk **24** (1980), no. 7, 588–591.

We compute below the groups $SUK_1(\tau, T)$ for division rings of noncommutative rational functions which are obtained as follows. Let A be a division ring; $\varphi_1, \varphi_2, \ldots, \varphi_n \in \operatorname{Aut} A$; $\{a_{ij}\}_{1 \le i \le j \le n} \subset A^*$. Consider the ring $F_{[n]} = A[x_1, \varphi_1, \ldots, x_n, \varphi_n, a_{ij}]$ of noncommutative polynomials in $x_1, \ldots, x_n$ connected with $\varphi_1, \ldots, \varphi_n$ respectively. (Note that $x_i a = a^{\varphi_i} x_i$ for an arbitrary $a \in A$ $(1 \le i \le n)$ and $x_j x_i = a_{ij} x_i x_j$, $i < j \le n$.) Suppose that $F_{[n]}$ is a right Ore ring and let $F_{(n)}$ be its division ring of (right) fractions. The division ring $F_{(n)}$ can be interpreted as follows. Consider the division ring $A_1 = A(x_1, \varphi_1) \subset F_{(n)}$. Then the last inclusion allows us to extend easily the automorphisms $\varphi_2, \ldots, \varphi_n$ to $\varphi_2^{(1)}, \ldots, \varphi_n^{(1)} \in \operatorname{Aut} A_1$ by setting $x_i^{\varphi_i^{(1)}} = a_{1i} x_1$, $1 < i \le n$. Next we consider the ring $A_1[x_2, \varphi_2^{(1)}] \subset F_{(n)}$ and its division ring of fractions $A_2 = A_1(x_2, \varphi_2^{(1)}) \subset F_{(n)}$. Suppose that the division ring $A_i \subset F_{(n)}$ is already constructed. As above, we set $A_{i+1} = A_i(x_{i+1}, \varphi_{i+1}^{(i)}) \subset F_{(n)}$, where A_{i+1} is the division ring of fractions of the ring $A_i[x_{i+1}, \varphi_{i+1}^{(i)}]$. Here, if $i + 1 < n$, we set $x_{i+1}^{\varphi_j^{(i+1)}} = a_{ij} x_{i+1}$, $i < j$, to obtain, taking into account that $A_{i+1} \subset F_{(n)}$, an extension of $\varphi_j^{(i)}$ to an automorphism $\varphi_j^{(i+1)} \in \operatorname{Aut} A_{i+1}$. Then it is not difficult to see that $F_{(n)} = A_n$.

We shall assume now that the division ring A_n is finite-dimensional. This condition implies that A is finite-dimensional, that the restriction G of the group $\widetilde{G} = \{\varphi_1, \ldots, \varphi_n\}$ to the center $Z(A)$ of the division ring A is finite, and that the "outer" orders r_i of the automorphisms $\varphi_i^{(i-1)}$, $1 \le i \le n$, are finite (here $\varphi_1^0 = \varphi_1$). Denote the field of G-invariants in $Z(A)$ by $Z(A)_G$. Then we have

LEMMA 1. *If $a \in A$, then*
$$\operatorname{Nrd}_{A_n}(a) = N_{Z(A)|Z(A)_G}(\operatorname{Nrd}_A(a))^{\lambda(A_n|Z(A))},$$
where $\lambda(A_n|Z(A)) = (\prod_{i=1}^n r_i)[Z(A) : Z(A)_G]^{-1}$.

The proof is not complicated.

From here on, the involutions τ_n of the division ring A_n satisfy the following condition: $A^{\tau_n} = A$, $A[x_i, \varphi_i]^{\tau_n} = A[x_i, \varphi_i]$, $i = 1, \ldots, n$. An easy argument shows that a suitable change of variables $x_1, \ldots, x_n$ allows us to assume without loss of generality that x^{τ_n} coincides with x_i, $-x_i$, or $-x_i + b_i$, where b_i is an element of the center of A_n ($\tau_n|_A = \tau$).

We assume now that τ_n is an involution of the second kind and that the A_n are not commutative (otherwise $SUK_1(\tau_n, A_n) = 1$). Here is the main lemma.

LEMMA 2. $\Sigma'_{\tau_n}(A_n) = (\Sigma'_{\tau_n}(A_n) \cap A)\Sigma_{\tau_n}(A_n)$.

PROOF (sketch). In the case when the index of A_n is odd, the lemma is a consequence of the corresponding statement for the group $SL(1, A_n)$. In the case of even index, we proceed by induction on the number n. Assume first that $n > 1$ and that the lemma holds in the case of division rings with less than n variables. In particular,

$$\Sigma'_{\tau_n}(A_n) = (\Sigma'_{\tau_n}(A_n) \cap A_1)\Sigma_{\tau_n}(A_n).$$

Then we prove the inclusion

$$(\Sigma'_{\tau_n}(A_n) \cap A_1) \subset (\Sigma'_{\tau_n}(A_n) \cap A)\Sigma_{\tau_n}(A_n).$$

Let $a \in \Sigma'_{\tau_n}(A_n) \cap A_1$ (without loss of generality we can assume that $a \in A[x_1, \varphi_1]$). Consider the element $u = (\prod_{g \in \widetilde{G}_1} a^g)^{\lambda(A_n|Z(A_1))}$ and its decomposition $u = P_1 P_2 \cdots P_\delta$ into irreducible factors in $A[x_1, \varphi_1]$ ($\widetilde{G}_1$ is the group generated by $\varphi_2^{(1)}, \ldots, \varphi_n^{(1)}$). By Lemma 1, $u \in \Sigma'_{\tau_1}(A_1)$. The last circumstance implies that there are $v, w \in A[x_1, \varphi_1]$ such that $\deg v = \deg w < \deg P_1$ and $wP_1 = P_c^{\tau_1 g} v$ for suitable $g \in \widetilde{G}_1$ and $1 \leq c \leq \delta$. In its turn, $wP_1 = P_c^{\tau_1 g} v$ implies that there is a $b \in A[x_1, \varphi_1]$ such that $\deg b < \deg a$, $a \equiv b \pmod{\Sigma_{\tau_n}(A_n)}$. Repetition of the argument a few times proves the inclusion $\Sigma'_{\tau_n}(A_n) \cap A_1 \subset (\Sigma'_{\tau_n}(A_n) \cap A)\Sigma_{\tau_n}(A_n)$. A similar argument establishes the validity of the lemma in the case $n = 1$, which completes the proof.

Denote by T the set $\{x_1^{\alpha_1}, \ldots, x_n^{\alpha_n} | \alpha_i = 0, 1\}$, and let $t \in T$. The leading term of the polynomial t^{τ_n} has the form $\mu_t t$ for a suitable $\mu_t \in A^*$. Below $\Sigma_t(A)$ denotes the normal subgroup of A^* generated by all elements a such that $\mu_t a^{\tau_n i_t} = a$. Let $A^{\widetilde{G}}$ be the normal subgroup of A^* generated by all elements a^{g-1}, where $g \in \widetilde{G}$, let Γ be the commutator subgroup of the subgroup of A^* with generators $x_1, \ldots, x_n$, and let $\Gamma_{\tau_n}(A_n)$ be the subgroup of A_n^* generated by all elements of A_n^* that are skew-symmetric with respect to τ_n.

The following holds:

LEMMA 3. *If $x_i^{\tau_n} \neq x_i$ for at least one $i \leq n$, $\operatorname{char} A \neq 2$, $\tau \neq \operatorname{id} A$, and the index of A_n is odd, then $\prod_{t \in T} \Sigma_t(A) A^{\widetilde{G}} \Gamma = \Gamma_{\tau_n}(A_n) \cap A$. Otherwise,*
$\prod_{t \in T} \Sigma_t(A) A^{\widetilde{G}} \Gamma = \Sigma_{\tau_n}(A_n) \cap A$.

PROOF (sketch). First we establish that every element of A_n has the form $P\lambda^{-1}$, where $P, \lambda \in F_{[n]}$, and λ is central in A_n. Then, if $a \in \Sigma_{\tau_n}(A_n) \cap A$ and $a = \sum_{i=1}^m P_i \lambda_i^{-1}$, $P_i, \lambda_i \in F_{[n]}$, λ_i are in the center of A_n, then the comparison of the leading coefficients in the right- and left-hand sides of the equality $a \prod_{i=1}^m \lambda_i = \prod_{i=1}^m P_i$ allows us to prove the inclusion $(\Sigma_{\tau_n}(A_n) \cap A) \subset \prod_{t \in T} \Sigma_t(A) A^{\widetilde{G}} \Gamma$. The inverse inclusion can be proved directly.

We fix an involution $\tau^* \in \{\tau\} \cup \{\tau\varphi_i | x_i^{\tau_n} = x_i\}_{i=1}^n$ and define the following groups:

$$\Sigma'_{\tau_n}(A_n)_G = \{a \in A^* | N_{Z(A)|Z(A)_G}(\mathrm{Nrd}_A(a))^{1-\tau^*} = 1\},$$

$$\tilde{\Sigma}'_{\tau_n}(A_n)_G = \Sigma'_{\tau_n}(A_n)_G \Sigma_{x_j}(A),$$

where $j = 1$ when $x_i^{\tau_n} = x_i$, $1 \le i \le n$, and j is the least number with $x_j^{\tau_n} \ne x_j$ otherwise;

$$\Sigma_{\tau_n}(A_n)_G = \Sigma_\tau(A) \prod_{i=1}^n \Sigma_{x_i}(A) A^{\tilde{G}},$$

$$SUK_1(\tau_n, A_n)_G = \tilde{\Sigma}'_{\tau_n}(A_n)_G / \Sigma_{\tau_n}(A_n)_G.$$

Set

$$PU(\tau^*, A) = (\mathrm{Ker}\, N_{Z(A)|Z(A)_G}|_{\mathrm{Nrd}_A(A^*)})\, \mathrm{Nrd}_A(\Sigma_{\tau_n}(A_n)_G)$$
$$\times (\mathrm{Nrd}_A(A^*) \cap Z(A)_{\tau^*}) / \mathrm{Nrd}_A(\Sigma_{\tau_n}(A_n)_G)(\mathrm{Nrd}_A(A^*) \cap Z(A_{\tau^*})).$$

The following statements describe interrelations among the groups defined above.

LEMMA 4. *The following sequence is exact*:

$$1 \to R \to \hat{H}^{-1}(G, \mathrm{Nrd}_A(A^*)) \to PU(\tau^*, A) \to 1,$$
$$R = (\mathrm{Ker}\, N_{Z(A)|Z(A)_G}|_{\mathrm{Nrd}_A(A^*)}) \cap \mathrm{Nrd}_A(\Sigma_{\tau_n}(A_n)_G)$$
$$\times (\mathrm{Nrd}_A(A^*) \cap Z(A)_{\tau^*}) / \mathrm{Nrd}_A(A^{\tilde{G}}).$$

LEMMA 5.
$$PU(\tau^*, A) \simeq (\mathrm{Ker}\, N_{Z(A)|Z(A)_G}|_{\mathrm{Nrd}_A(A^*)})^{1-\tau^*}$$
$$\times \mathrm{Nrd}_A(\Sigma_{\tau_n}(A_n)_G)^{1-\tau^*} / \mathrm{Nrd}_A(\Sigma_{\tau_n}(A_n)G)^{1-\tau^*}.$$

For an arbitrary $a \in \tilde{\Sigma}'_{\tau_n}(A_n)$ we set

$$\delta_n(a) = \mathrm{Nrd}_A(a)^{1-\tau^*} \mathrm{Nrd}_A(\Sigma_{\tau_n}(A_n)_G)^{1-\tau^*}.$$

LEMMA 6. δ_n *is an epimorphism from* $\tilde{\Sigma}'_{\tau_n}(A_n)_G$ *to*

$$(\mathrm{Ker}\, N_{Z(A)|Z(A)_G}|_{\mathrm{Nrd}_A(A^*)})^{1-\tau^*} \mathrm{Nrd}_A(\Sigma_{\tau_n}(A_n)_G)^{1-\tau^*} / \mathrm{Nrd}_A(\Sigma_{\tau_n}(A_n)_G)^{1-\tau^*}$$

with kernel $\Sigma'_{\tau^*}(A)\Sigma_{\tau_n}(A_n)_G$.

From Lemmas 1–6 the main computational theorem follows easily:

THEOREM. *The following sequences are exact*:

$$1 \to SUK_1(\tau_n, A_n)_G/M \to SUK_1(\tau_n, A_n) \to Z/\mu Z \to 1, \qquad (1)$$

where $\mu | \lambda(A_n | Z(A))$,

$$M \simeq \frac{(\Sigma'_{\tau_n}(A_n)_G \cap \prod_{t \in T \setminus U} \Sigma_\tau(A)\Gamma)}{(\prod_{t \in T \setminus U} \Sigma_t(A)\Gamma \cap (\prod_{t \in U} \Sigma_t(A))\Sigma_\tau(A) A^{\widetilde{G}})},$$

and $U = \{x_1, \ldots, x_n\} \cup 1$;

$$1 \to SUK_1(\tau^*, A)/N \to SUK_1(\tau_n, A_n)_G \to PU(\tau^*, A) \to 1, \qquad (2)$$

where $N = (\Sigma'_{\tau^*}(A) \cap \Sigma_{\tau_n}(A_n)_G)/\Sigma_{\tau^*}(A)$; and

$$1 \to R \to \widehat{H}^{-1}(G, \mathrm{Nrd}_A(A^*)) \to PU(\tau^*, A) \to 1 \qquad (3)$$

(see Lemma 4).

REMARK. In various particular cases, from the previous results one can derive more explicit formulas (cf., e.g., [5] about the case $n = 1$, $x_1^{\tau_1} = x_1$).

Bibliography

1. V. P. Platonov, *The Tannaka-Artin problem, and reduced K-theory*, Izv. Akad. Nauk SSSR Ser. Mat. **40** (1976), no. 2, 227–261; English transl. in Math. USSR-Izv. **10** (1976).
2. V. P. Platonov and V. I. Yanchevskiĭ, *Stability in reduced K-theory*, Dokl. Akad. Nauk SSSR **242** (1978), 769–772; English transl. in Soviet Math. Dokl. **19** (1978).
3. ____, *SK_1 for division rings of noncommutative rational functions*, Dokl. Akad. Nauk SSSR **249** (1979), 1064–1068; English transl. in Soviet Math. Dokl. **20** (1979).
4. V. I. Yanchevskiĭ, *Reduced unitary K-theory and division algebras over Henselian discretely valued fields*, Izv. Akad. Nauk SSSR Ser. Mat. **42** (1978), no. 4, 879–918; English transl. in Math. USSR-Izv. **13** (1979).
5. ____, *Reduced Whitehead unitary groups of skew-fields of noncommutative rational functions*, Zap. Nauchn. Sem. Leningrad. Otdel. Mat. Inst. Steklov. (LOMI) **94** (1979), 142–148; English transl. in J. Soviet Math. **19** (1982), no. 1.

Institute of Mathematics
Academy of Sciences of the Belorussian SSR

Received 23/OCT/79

Translated by L. N. VASERSTEIN

Rational Splitting Fields of Simple Algebras and Unirationality of Conic Bundles

UDC 512.552 + 512.774

I. I. VORONOVICH AND V. I. YANCHEVSKIĬ

Abstract. Central simple algebras over the rational function field $k(x)$ are considered, where k is an infinite algebraic extension of a finite field F_q. It is proved with some restriction on an index of a given algebra that it possesses the splitting field, which is a rational function field over k. Thus, if k is such a field, then every k-defined conic bundle over the k-rational curve is k-unirational.

Let k be a field, $k(x)$ a purely transcendental extension of k of degree 1, and $V_{k(x)}$ the set of places of $k(x)$ that are trivial on k. For an element $v \in V_{k(x)}$ let $k(x)_v$ denote the completion of $k(x)$ at the place v. We recall the following definition from [1].

DEFINITION. A finite-dimensional central simple algebra A over the field $k(x)$ is said to have a k-place if there exists a place $v \in V_{k(x)}$ of degree 1 such that the algebra $A \otimes_{k(x)} k(x)_v$ is the full matrix algebra over the field $k(x)_v$.

In connection with the problem of k-unirationality of pencils of conics (see [2]) the following conjecture was formulated in [1]: Every finite-dimensional simple central algebra over the field $k(x)$ having a k-place has a splitting field that is a purely transcendental extension of k of degree 1. There it was also shown that the conjecture is valid in the case when k is a Henselian field.

In the present note it is shown that the conjecture formulated is valid for central simple algebras over $k(x)$ in the case when k is an infinite algebraic extension of a finite field and contains some primitive roots of unity whose degrees are connected with the indices of the algebras under consideration. We remark that the condition of the conjecture is satisfied automatically in

---

1991 *Mathematics Subject Classification.* Primary 11R52, 16H05.
Translation of Dokl. Akad. Nauk BSSR **30** (1986), no. 4, 293–296.

this case, since for any central simple algebra A over $k(x)$ almost all places $v \in V_{k(x)}$ of degree 1 are k-places of A in the sense of the definition given.

We hold to the notation and conventions in [1]. In particular, all the algebras under consideration are finite-dimensional. For brevity we shall use the phrase "field rational over k" instead of "purely transcendental extension of k of degree 1".

Our basic result is the following

THEOREM. *Let k be an infinite algebraic extension of a finite field F_q of q elements, and let A be a central simple algebra over $k(x)$ of index $p_1^{\alpha_1} \cdots p_t^{\alpha_t} \cdot \operatorname{char} k^\mu$, $\alpha_1, \ldots, \alpha_t > 0$, $\mu \geq 0$. Assume that k contains a primitive root of unity of degree $p_1 \cdots p_t$. Then A has a splitting field rational over k.*

We give two important corollaries to this theorem.

COROLLARY 1. *Suppose that F_q is a finite field, and A is a simple algebra of prime power index with center rational over F_q. Then A has a splitting field that is rational over a finite extension of F_q of degree relatively prime to the index of A.*

COROLLARY 2. *If k is an infinite algebraic extension of a finite field F_q (char $F_q \neq 2$), then any conic bundle defined over k is k-unirational. Therefore, every conic bundle defined over F_q is L-unirational for some finite extension $L|F_q$ of odd degree.*

We precede the proof of the theorem with some remarks and auxiliary assertions.

REMARK 1. To prove the theorem it suffices to confine ourselves to algebras of prime power index. Indeed, assume that the theorem is proved for such algebras, and let A be a simple central algebra over the field $k(x)$. The algebra A is similar to the tensor product $A_{p_1} \otimes_{k(x)} \cdots \otimes_{k(x)} A_{p_s}$, where A_{p_i} is a division algebra of index a power of p_i, and $p_1, \ldots, p_s$ are distinct prime numbers. Now if $k(z_s)$ is a splitting field of the algebra A_{p_s}, then the algebra $A \otimes_{k(x)} k(z_s)$ is similar to a tensor product $B_1 \otimes_{h(x)} \cdots \otimes_{k(x)} B_r$ of division algebras of prime power indices, where $r < s$. A simple induction argument then shows that the theorem is valid also for A.

REMARK 2. In proving the theorem for an arbitrary algebra of prime power index it suffices to consider the case when its index is relatively prime to the characteristic of k. Indeed, suppose that the index of A is p^n, where $p = \operatorname{char} k$. Let $x = z^{p^n}$ and consider the rational function field $k(z)$. Then $k(z)$ is the desired splitting field of the algebra A. Indeed, A is similar to the crossed product $(L(x), G, a_{S,T}(x))_{k(x)}$, where, as usual, $L|k$ is a Galois extension, G is the Galois group of $L(x)$ over $k(x)$, and $a_{S,T}(x)$ is the system of factors on G with values in the field $L(x)$. Then

$$(L(x), G, a_{S,T}(x))_{k(x)} \otimes_{k(x)} k(z)$$
$$= (L(z), G, a_{S,T}(z^{p^n}))_{k(z)} = (L(z), G, a_{S,T}(z)^{p^n})_{k(z)}$$
$$\sim (L(z), G, a_{S,T}(z))_{k(z)}^{p^n} \sim 1,$$

which is what was required.

REMARK 3. Let A be a primary algebra of index p^n over $k(x)$, $n > 0$, $\operatorname{char} k \neq p$, and suppose that the field k contains a primitive pth root of unity ξ_p. Then $k^p \neq k$. Indeed, there exists a finite extension $L|k$ of degree p^l such that $L(x)$ is a splitting field of A. If now $k^p = k$, then since $\xi_p \in k$ and $p \neq \operatorname{char} k$, there are no nontrivial extensions of k of degree p. Consequently, $l = 0$, which contradicts the nontriviality of A. In what follows, γ denotes a fixed element of $k^* \backslash k^{*p}$.

The following lemmas are needed in the proof of the theorem.

LEMMA 1. *Let K be the field of quotients of an integral domain A, and $f(x) \in A[x]$ a unitary polynomial. Then the existence of divisors of zero in the ring $A[x]/(f(x))$ implies that the ring $K[x]/(f(x))$ also has divisors of zero.*

The proof of the lemma is a simple exercise in commutative algebra, and we omit it.

Let $g_1(x), \ldots, g_r(x) \in k[x]$ be a system of distinct irreducible polynomials of positive degrees. Denote by k_0 the minimal subfield in k containing γ and all the coefficients of the polynomials $g_1(x), \ldots, g_r(x)$. If, as usual, F_q is a field of q elements, then denote by S the set $\{q | k_0 \subset F_q \subset k\}$. Let p be a prime number different from the characteristic of k. For a tuple $(\alpha_1, \ldots, \alpha_r)$ with $\alpha_i \in \{0, 1, \ldots, p-1\}$, $i = 1, \ldots, r$, we define a function $T_{\alpha_1, \ldots, \alpha_r}(q)$ on S by $T_{\alpha_1, \ldots, \alpha_r}(q) = \operatorname{card}\{a \in F_q | g_i(a) \in \gamma^{\alpha_i} F_q^{*p}, i = 1, \ldots, r\}$.

LEMMA 2. *In the notation above,*

$$T_{\alpha_1, \ldots, \alpha_r}(q) = q/p^r + \mathcal{O}(q^{1/2}). \tag{1}$$

PROOF. Let $\bar{k}$ be the algebraic closure of the field k. We show that the k_0-defined affine curve $\Gamma \subset A_{\bar{k}}^{r+1}$, given by the system of equations

$$g_1(x) = \gamma^{\alpha_1} y_1^p,$$
$$\ldots\ldots\ldots\ldots\ldots$$
$$g_r(x) = \gamma^{\alpha_r} y_r^p,$$

is irreducible. For this it suffices to show that the ring $\bar{k}[x, y_1, \ldots, y_r]/(y_1^p - g_1(x), \ldots, y_r^p - g_r(x))$ is an integral domain for any $r \geq 1$. If $r = 1$, then by Lemma 1 it suffices to show that $\bar{k}(x)[y_1]/(y_1^p - g_1(x))$ is an integral domain. Since the polynomial $g_1(x)$ is irreducible in the ring $k[x]$, all the roots are distinct. Consequently, $g_1(x) \notin \bar{k}(x)^{*p}$, and hence the polynomial $y_1^p - g_1(x)$ is irreducible in the ring $\bar{k}(x)[y_1]$. Thus, the principal ideal $(y_1^p - g_1(x))$ of the ring $\bar{k}(x)[y_1]$ is simple, which is what was required. Suppose now that our assertion is valid for $r = l \geq 1$, i.e., the ring $\bar{k}[x, \sqrt[p]{g_1(x)}, \ldots, \sqrt[p]{g_l(x)}]$ is an integral domain. Denote by K its field of quotients. Then again by Lemma 1, it suffices to show that the ring $K[y_{l+1}]/(y_{l+1}^p - g_{l+1}(x))$ is an integral domain. Assume not. Then the polynomial $y_{l+1}^p - g_{l+1}(x)$ factors into linear factors in the ring $K[y_{l+1}]$, i.e., the field $\bar{k}(x)(\sqrt[p]{g_{l+1}})$ is contained in K. It is clear that K is an Abelian extension of $\bar{k}(x)$ of exponent p. According to the theory of Kummer, $g_{l+1} = g_1^{\beta_1} \cdots g_l^{\beta_l} h^p$, where $h \in k[x]$, but this cannot be true, because the polynomials $g_i(x)$, $i = 1, \ldots, l+1$, are relatively prime. Thus, the curve Γ is absolutely irreducible.

Let $\bar{\Gamma}$ be the closure of Γ in the projective space $P_{\bar{k}}^{r+1}$, which contains Γ. It is not hard to see that the curve $\bar{\Gamma}$ has a k_0-defined normalization $\bar{\Gamma}^\nu$ together with a k_0-birational regular mapping $\nu: \bar{\Gamma}^\nu \to \bar{\Gamma}$. Therefore, there exists a k_0-birational isomorphism $\bar{\Gamma}^\nu \to \Gamma$ of the curves that is k_0-defined and biregular on certain open subsets $U' \subset \bar{\Gamma}^\nu$ and $U \subset \Gamma$. This isomorphism establishes a bijective correspondence between the F_q-places of the sets U' and U, where F_q is an arbitrary finite field such that $k_0 \subset F_q \subset k$. For each such field let N_q (respectively, N_q') denote the number of F_q-places of the curve Γ (respectively, of the curve $\bar{\Gamma}^\nu$). Since the complements $\Gamma \setminus U$ and $\bar{\Gamma}^\nu \setminus U'$ are finite, it follows that $N_q = N_q' + \mathcal{O}(1)$. In view of the normality (and hence smoothness) of the projective curve $\bar{\Gamma}^\nu$ the number N_q' satisfies the Weil inequality $|N_q' - (q+1)| \leq 2g\sqrt{q}$ [3], where g is the genus of Γ. This implies that $N_q' = q + \mathcal{O}(q^{1/2})$, and hence

$$N_q = q + \mathcal{O}(q^{1/2}). \qquad (2)$$

Now let Γ_{F_q} denote the set of F_q-places of the curve Γ (then $N_q = \operatorname{card} \Gamma_{F_q}$). We consider the projection $\varphi: \Gamma_{F_q} \to A_{F_q}^1$ on the first coordinate: $\varphi(x, y_1, \ldots, y_r) = x$. Each point $x \in \varphi(\Gamma_{F_q})$, with the exception of the points $x \in F_q$ for which $g_1(x) \cdots g_r(x) = 0$, has inverse image $\varphi^{-1}(x)$ consisting of p^r distinct points of the curve Γ, because F_q contains a primitive pth root of unity. This means that $N_q = p^r \cdot T_{\alpha_1, \ldots, \alpha_r}(q) + t$, where t is the number of points of the set Γ_{F_q} such that $g_1(x) \cdots g_r(x) = 0$. Therefore,

$$N_q = p^r \cdot T_{\alpha_1, \ldots, \alpha_r}(q) + \mathcal{O}(1). \qquad (3)$$

The equality (1) now follows from the equalities (2) and (3). The lemma is proved.

COROLLARY. *If $g_1(x), \ldots, g_r(x)$ are distinct irreducible polynomials of positive degrees over the field k, then there exists an element $\alpha \in k$ such that $g_1(\alpha) \in k^* \setminus k^{*p}, \ldots, g_r(\alpha) \in k^* \setminus k^{*p}$.*

Indeed, since, for example, $T_{1,\ldots,1}(q) = q/p^r + \mathcal{O}(q^{1/2})$, it follows that $T_{1,\ldots,1}(q) > 0$ for sufficiently large q, i.e., the field $F_q \subset k$ contains an element α such that $g_i(\alpha) \in \gamma k^{*p}$, $i = 1, \ldots, r$. In particular, $g_i(\alpha) \in k^* \setminus k^{*p}$, which is what was required.

LEMMA 3. *For any set of distinct unitary irreducible polynomials over k and any natural number n there is a polynomial $\varphi(x)$ over k such that all the $f_i(\varphi(x))$ are irreducible and $\deg f_i(\varphi(x))$ is divisible by p^n, $i = 1, \ldots, r$.*

It is not hard to see that the proof of the general assertion can be reduced to the case $n = 1$. Let $\varphi(x) = x^p - a$, where $a \in k$. Let $f_i = \prod_j (x - \alpha_{ij})$, $i = 1, \ldots, r$, and $g_i(x) = \prod_j (x + \alpha_{ij})$. We show that for the polynomials $f_i(\varphi(x))$ to be irreducible over k it suffices to take the element a such that $g_i(a) \in k^* \setminus k^{*p}$. Indeed,

$$f_i(\varphi(x)) = \prod_j (\varphi(x) - \alpha_{ij}) = \prod_j (x^p - (a + \alpha_{ij})),$$

and it is clear that the polynomial $f_i(\varphi(x))$ is irreducible over k if and only if the polynomial $x^p - (a + \alpha_{ij})$ is irreducible over the field $k(\alpha_{ij})$, which is equivalent to the condition $a + \alpha_{ij} \in k(\alpha_{ij})^* \setminus k(\alpha_{ij})^{*p}$. The last condition will hold if $g_i(a) \in k^* \setminus k^{*p}$. But the existence of an element $a \in k$ with this property is ensured by the corollary to Lemma 2. The lemma is proved.

PROOF OF THE THEOREM. By Remarks 1 and 2, it can be assumed that the index of A is equal to p^n and $p \neq \operatorname{char} k$. Then A is similar to a cyclic algebra $(L(x), \prod_i f_i^{\alpha_i}(x))_{k(x)}$, where L is an extension of k of degree p^t, $t \geq n$, and the $f_i(x)$ are distinct unitary irreducible polynomials. According to Lemma 3, there exists a polynomial $\varphi(x) \in k[x]$ such that the polynomials $f_i(\varphi(x))$ are irreducible, and their degrees are divisible by p^t. We set $x = \varphi(z)$ and show that the field $k(z)$ splits the algebra A. Indeed,

$$A \otimes_{k(x)} k(z) \sim \left(L(x), \prod_i f_i(x)^{\alpha_i}\right)_{k(x)} \otimes_{k(x)} k(z)$$

$$= \left(L(z), \prod_i f_i(\varphi(z))^{\alpha_i}\right)_{k(z)} \sim \prod_i (L(z), f_i(\varphi(z)))_{k(z)}^{\alpha_i}.$$

It now suffices to show that for each i the algebra $(L(z), f_i(\varphi(z)))_{k(z)}$ is a matrix algebra. Since $\deg f_i(\varphi(z))$ is divisible by p^t, we have a tower of fields $k(z) \subset L(z) \subset P_i(z)$, where P_i is a splitting field of the polynomial $f_i(\varphi(z))$, $P_i = k(\beta_i)$ for some root β_i of $f(\varphi(z))$. Since the norm is multiplicative,

$$f_i(\varphi(z)) = N_{P_i(z)|k(z)}(z - \beta_i) = N_{L(z)|k(z)}(N_{P_i(z)|L(z)}(z - \beta_i)).$$

But $N_{P_i(z)|L(z)}(z - \beta_i) \in L(z)$, so that $f_i(\varphi(z))$ is the norm over $k(z)$ of an element in $L(z)$, which is what was required. Accordingly, $k(z)$ is the desired field. The theorem is proved.

In conclusion we remark that in connection with Corollary 1 we can formulate the following weakened variant of the conjecture in [1].

Let k be an arbitrary field, and A a finite-dimensional central simple algebra of index n over the field $k(x)$. Then there exists an extension L of k such that $\mathrm{ind}(A \otimes_{k(x)} L(x)) = n$, and A is split by some field that is rational over L.

Bibliography

1. V. I. Yanchevskiĭ, *K-unirationality of conic bundles and splitting fields of simple central algebras*, Dokl. Akad. Nauk BSSR **29** (1985), no. 12, 1061–1064. (Russian)
2. V. A. Iskovskikh, *Rational surfaces with a pencil of rational curves*, Mat. Sb. **74** (1967), 608–638; English transl. in Math. USSR-Sb. **3** (1967).
3. Robin Hartshorne, *Algebraic geometry*, Springer-Verlag, Berlin and New York, 1977.

Institute of Mathematics
Academy of Sciences of the Belorussian SSR

Received 26/APR/85

Translated by H. H. McFADEN

Crossed Products of Simple Algebras and Their Automorphism Groups

UDC 512.552.33

V. V. KURSOV AND V. I. YANCHEVSKIĬ

Abstract. A generalization of the classical theory of crossed products is presented.

Crossed products of fields and their Galois groups, which were first used by Noether (see [8]) and Brauer (see [4]), remain up to the present time one of the most important constructions in the theory of simple finite-dimensional algebras. This construction was later generalized repeatedly from different points of view by Albert (see [1]), Jacobson [5], [6], Teichmüller [3], Tikhomirov [14], and others ([7], [9], and [13]).

Investigations of Henselian division algebras (see [10]–[12]) stimulated the interest of the second author of this paper in constructions of crossed products of simple algebras with special groups of their automorphisms. The general construction of a crossed product of an arbitrary associative ring and a semigroup is due to Bovdi (see [2]), and he also obtained a number of results on these crossed products. In the present article we are interested in the more special situation of crossed products of simple finite-dimensional associative central algebras K over $Z(K)$, and of their groups G of automorphisms with restriction to the center $Z(K)$ forming finite groups first considered by Teichmüller (see [3]). In this situation it is possible to get a generalization of basic results in the theory of classical crossed products of Noether. Our main goal is to formulate these results. We remark that the construction to be investigated by us is important in the theory of finite-dimensional division algebras, since it is known that every Henselian division algebra with weak nontrivial ramification can be obtained by means of such a construction.

Let K be a simple algebra over its center Z, and let $\text{Aut}(K)$ ($\text{Int}(K)$)

1991 *Mathematics Subject Classification.* Primary 16A40; Secondary 12E15, 16A39, 16A74.
Translation of Dokl. Akad. Nauk BSSR **32** (1988), no. 9, 777–780.

©1992 American Mathematical Society
0065-9290/92 $1.00 + $.25 per page

be the automorphism group of K (the subgroup of inner automorphisms of K). We consider a group $G \subset \mathrm{Aut}(K)$ such that its restriction to Z gives a finite group $\widetilde{G}$. Denote by $Z_{\widetilde{G}}$ the subfield of invariants of $\widetilde{G}$ in Z. For each $\sigma \in \widetilde{G}$ we fix $\overline{\sigma} \in G$ such that $\overline{\sigma}|_Z = \sigma$. Let $f \colon \widetilde{G} \times \widetilde{G} \to K^*$ be a given mapping, where K^* denotes the group of invertible elements of the algebra K.

DEFINITION 1 (see [2]). A pair $(f, \overline{\sigma}|\sigma \in \widetilde{G})$ is called a generalized system of factors with values in K if for any $\sigma, \tau, \nu \in \widetilde{G}$ and any $k \in K$

$$f(\sigma\tau, \nu)f(\sigma, \tau)^{\overline{\nu}} = f(\sigma, \tau\nu)f(\tau, \nu),$$
$$k^{\overline{\tau\nu}} = f(\tau, \nu)k^{\overline{\tau}\,\overline{\nu}}f(\tau, \nu)^{-1}.$$

DEFINITION 2 (see [2]). Let $(f, \overline{\sigma}|\sigma \in \widetilde{G})$ be a generalized system of factors. The ring $A = (K, \overline{\sigma}, f)_{\sigma \in \widetilde{G}}$ is called the crossed product of K and G with respect to f if $A = \{\sum_{\sigma \in \widetilde{G}} x_\sigma k_\sigma | k_\sigma \in K\}$, with equality and addition in A defined componentwise and multiplication defined by the conditions

$$kx_\sigma = x_\sigma k^{\overline{\sigma}} \quad \text{for all } \sigma \in \widetilde{G} \text{ and } k \in K,$$
$$x_\sigma x_\tau = x_{\sigma\tau} f(\sigma, \tau) \quad \text{for any } \sigma, \tau \in \widetilde{G}.$$

The algebra K will be called the frame of the crossed product, and the K-basis $\{x_\sigma\}_{\sigma \in \widetilde{G}}$ is called the standard basis of A.

The following assertion characterizes the structure of a crossed product.

THEOREM 1. *A is an associative ring with identity for which the following assertions are valid*:

(i) *A contains a subring isomorphic to K*;

(ii) *for any $\sigma \in \widetilde{G}$ the basis element x_σ is invertible in A and induces the automorphism $\overline{\sigma}$ in inner fashion, i.e., $x_\sigma^{-1} k x_\sigma = k^{\overline{\sigma}}$ for any $k \in K$*;

(iii) *the centralizer of K in A coincides with Z, and the center of the whole crossed product is equal to $Z_{\widetilde{G}}$*;

(iv) *A is a simple ring*.

DEFINITION 3. Two generalized systems of factors $(f, \overline{\sigma}|\sigma \in \widetilde{G})$ and $(g, \widetilde{\sigma}|\sigma \in \widetilde{G})$ are said to be equivalent if there exists a function $t \colon \widetilde{G} \to K^*$ such that:

1) $\widetilde{\sigma} = \overline{\sigma} \cdot i_{t_\sigma}$ for any $\sigma \in \widetilde{G}$, where i_{t_σ} denotes the inner automorphism of K induced by the element t_σ;

2) $g(\sigma, \tau) = t_{\sigma\tau} f(\sigma, \tau) t_\sigma^{-\overline{\tau}} t_\tau^{-1}$ for any $\sigma, \tau \in \widetilde{G}$.

This definition clearly agrees with the classical definition of equivalent systems of factors. All the crossed products considered by us below contain finite-dimensional frames. The following assertion gives necessary and sufficient conditions for an isomorphism of crossed products with the same frame.

THEOREM 2. *The crossed products* $(K, \overline{\sigma}, f)_{\sigma \in \widetilde{G}}$ *and* $(K, \check{\sigma}, g)_{\sigma \in \widetilde{G}}$ *are isomorphic over* $Z_{\widetilde{G}}$ *if and only if the generalized systems of factors* $(f, \overline{\sigma} | \sigma \in \widetilde{G})$ *and* $(g, \check{\sigma} | \sigma \in \widetilde{G})$ *are equivalent.*

For an arbitrary algebra A with multiplication "$\cdot$", let A^{op} denote the algebra opposite to it:

$$A^{\mathrm{op}} = \{a \in A | a \times b \stackrel{\mathrm{df}}{=} b \cdot a \text{ for all } a, b \in A\}.$$

We consider an arbitrary crossed product $(K, \overline{\sigma}, f)_{\sigma \in \widetilde{G}}$ and an anti-automorphism $\Psi: K \to K^{\mathrm{op}}$ of Z-algebras such that $\Psi(k) = k$ for any $k \in K$. Then $\Psi^{-1} \overline{\sigma} \Psi \in \mathrm{Aut}(K^{\mathrm{op}})$ for arbitrary $\sigma \in \widetilde{G}$, and it is easy to verify that $(f^{-\Psi}, \Psi^{-1} \overline{\sigma} \Psi | \sigma \in \widetilde{G})$ is a generalized system of factors, where $f^{-\Psi}: \widetilde{G} \times \widetilde{G} \to K^{\mathrm{op}*}$ is defined by $f^{-\Psi}(\sigma, \tau) \stackrel{\mathrm{df}}{=} [f(\sigma, \tau)^{-1}]^{\Psi}$ for all $\sigma, \tau \in \widetilde{G}$.

THEOREM 3. *There is an isomorphism of crossed products*

$$(K, \overline{\sigma}, f)^{\mathrm{op}}_{\sigma \in \widetilde{G}} \cong (K^{\mathrm{op}}, \Psi^{-1} \overline{\sigma} \Psi, f^{-\Psi})_{\sigma \in \widetilde{G}}.$$

Now suppose that K and L are simple algebras with center Z, where $(K, \overline{\sigma}, f)_{\sigma \in \widetilde{G}}$ and $(L, \check{\sigma}, g)_{\sigma \in \widetilde{G}}$ are crossed products, and the automorphism groups of K and L induce on Z one and the same group $\widetilde{G} = \mathrm{Gal}(Z | Z_{\widetilde{G}})$. We consider the tensor product $K \otimes_Z L$ and define a generalized system of factors $(f \otimes g, \overline{\sigma} \otimes \check{\sigma} | \sigma \in \widetilde{G})$ as follows:

$$(\overline{\sigma} \otimes \check{\sigma})(k \otimes l) = k^{\overline{\sigma}} \otimes l^{\check{\sigma}} \quad \text{for any } \sigma \in \widetilde{G}, \ k \in K, \ l \in L,$$

$$f \otimes g: \widetilde{G} \times \widetilde{G} \to (K \otimes_Z L)^\times, \quad (f \otimes g)(\sigma, \tau) = (f(\sigma, \tau) \otimes 1)(1 \otimes g(\sigma, \tau))$$

for any $\sigma, \tau \in \widetilde{G}$.

Recall that two finite-dimensional simple algebras A and B with the same center are similar, $A \sim B$, if they represent a single element of the Brauer group.

THEOREM 4. *If* $(K, \check{\sigma}, f)_{\sigma \in \widetilde{G}}$ *and* $(L, \check{\sigma}, g)_{\sigma \in \widetilde{G}}$ *are crossed products, then*

$$(K, \overline{\sigma}, f)_{\sigma \in \widetilde{G}} \otimes_{Z_{\widetilde{G}}} (L, \check{\sigma}, g)_{\sigma \in \widetilde{G}} \sim (K \otimes_Z L, \overline{\sigma} \otimes \check{\sigma}, f \otimes g)_{\sigma \in \widetilde{G}}.$$

Assume that $(K, \overline{\sigma}, f)_{\sigma \in \widetilde{G}}$ is an arbitrary crossed product and L is an extension of the field $Z_{\widetilde{G}}$ such that the center Z of the algebra K and the field L are contained in a single ambient field. Let $\mathrm{Gal}(ZL | L)$ be the Galois group of the extension $ZL | L$, where ZL denotes the composite of the fields Z and L, $N \stackrel{\mathrm{df}}{=} L \cap Z$, and $\varphi: \mathrm{Gal}(ZL | L) \to \mathrm{Gal}(Z | N)$ is the group isomorphism obtained by restriction of automorphisms. We consider the crossed product

$$(K \otimes_Z ZL, \check{\sigma}, \tilde{f})_{\varphi^{-1}(\sigma) \in \mathrm{Gal}(ZL | L)}$$

with the generalized system of factors constructed according to the rules

$$\tilde{\sigma} = \overline{\sigma} \otimes \varphi^{-1}(\sigma) \quad \text{for each } \sigma \in \text{Gal}(Z|N),$$

$$\tilde{f}(\varphi^{-1}(\sigma), \varphi^{-1}(\tau)) = f(\sigma, \tau) \quad \text{for all } \sigma, \tau \in \text{Gal}(Z|N).$$

THEOREM 5. *The algebra* $(K, \overline{\sigma}, f)_{\sigma \in \widetilde{G}} \otimes_{Z_{\widetilde{G}}} L$ *is similar to the crossed product*

$$(K \otimes_Z ZL, \tilde{\sigma}, \tilde{f})_{\varphi^{-1}(\sigma) \in \text{Gal}(ZL|L)},$$

where $\sigma \in \text{Gal}(Z|N)$.

It turns out that, up to similarity, consideration of crossed products reduces to the case of crossed products whose frames are division algebras. Let $(K, \overline{\sigma}, f)_{\sigma \in \widetilde{G}}$ be a crossed product, where K is a finite-dimensional simple central algebra over the field Z. By Wedderburn's theorem, $K = M_r(T)$, where T is a finite-dimensional central division algebra over Z. Further, let $\tilde{\sigma} \in \text{Aut}(M_r(T))$ be such that $\tilde{\sigma}(T) = T$, and assume the relation $\tilde{\sigma}|_Z = \overline{\sigma}|_Z = \sigma$ for the restrictions.

Then we have

THEOREM 6. *Every crossed product* $(K, \overline{\sigma}, f)_{\sigma \in \widetilde{G}}$ *for which* $K = M_r(T)$ *and* T *is a finite-dimensional division algebra over its center* Z *is similar to some crossed product* $(T, \tilde{\sigma}|_T, x)_{\sigma \in \widetilde{G}}$, *where* $\tilde{\sigma} \in \text{Aut}(M_r(T))$ *is such that* $\tilde{\sigma}(T) = T$, $\tilde{\sigma}|_Z = \overline{\sigma}|_Z = \sigma$ *for any* $\sigma \in \widetilde{G}$, *and the generalized system of factors* $(x, \tilde{\sigma}|_T | \sigma \in \widetilde{G})$ *can be explicitly expressed in terms of* $(f, \overline{\sigma} | \sigma \in \widetilde{G})$.

In conclusion we consider the question of the connection between the construction we are considering and the classical crossed product of a field and its Galois group.

Consider a crossed product $(K, \overline{\sigma}, f)_{\sigma \in \widetilde{G}}$ with frame K that is a finite-dimensional central division algebra over the field Z. Generally speaking, there need not exist in the division algebra K a maximal subfield that is normal over $Z_{\widetilde{G}}$. However, there exists a natural number n such that the algebra $M_n(K)$ contains a subfield $R \supset Z \supset Z_{\widetilde{G}}$ that is an extension of the Galois field $Z_{\widetilde{G}}$ and such that $[R:Z] = nm$, where m is the index of the division algebra K. Let us consider the crossed product $(M_n(Z), \sigma_0, 1)_{\sigma \in \widetilde{G}}$, where an automorphism $\sigma_0 \in \text{Aut}(M_n(Z))$ means the application of σ to the elements of a matrix over Z. Then by Theorem 4, $(K, \overline{\sigma}, f)_{\sigma \in \widetilde{G}}$ is similar to $(K \otimes_Z M_n(Z), \overline{\sigma} \otimes \sigma_0, f \otimes 1)_{\sigma \in \widetilde{G}}$. By the choice of the field R with equality of the dimensions taken into account, we get an isomorphism of $Z_{\widetilde{G}}$-algebras

$$(K \otimes_Z M_n(Z), \overline{\sigma} \otimes \sigma_0, f \otimes 1)_{\sigma \in \widetilde{G}} \cong (R, \text{Gal}(R|Z_{\widetilde{G}}), a_{\tilde{\sigma}, \tilde{\tau}}),$$

where the last algebra is the classical crossed product of the field R and its Galois group. What is more, it is not hard to establish the dependence between systems of factors of the indicated crossed products. Namely, there exists a function

$$\mu: \mathrm{Gal}(R|Z_{\widetilde{G}}) \to (K \otimes_Z M_n(Z))^*$$

such that for any $\tilde{\sigma}, \tilde{\tau} \in \mathrm{Gal}(R|Z_{\widetilde{G}})$ such that $\tilde{\sigma}|_Z = \sigma$ and $\tilde{\tau}|_Z = \tau$

$$a_{\tilde{\sigma},\tilde{\tau}} = \mu_{\widetilde{\sigma\tau}}(f(\sigma,\tau) \otimes 1)\mu_{\tilde{\sigma}}^{\overline{\tau} \otimes \tau_0}. \qquad (*)$$

THEOREM 7. *Every crossed product* $(K, \overline{\sigma}, f)_{\sigma \in \widetilde{G}}$ *whose frame is a finite-dimensional central division algebra* K *is similar to some classical crossed product of a field and its Galois group, and the connection between the corresponding systems of factors is given by the relation* $(*)$.

We remark that in the case of a field the construction we have studied gives the usual crossed product of a field and its Galois group. Thus, all the results presented above contain properties of the usual crossed products as a special case.

Added in print

After the article went to print the author learned some of the results had been obtained by Professor J.-P. Tignol (see [15]).

BIBLIOGRAPHY

1. A. A. Albert, *Non-associative algebras* II. *New simple algebras*, Ann. of Math. (2) **43** (1942), 708–723.
2. A. A. Bovdi, *Crossed products of a semigroup and a ring*, Dokl. Akad. Nauk SSSR **137** (1961), 1267–1269; English transl. in Soviet Math. Dokl. **2** (1961).
3. Oswald Teichmüller, *Über die sogenannte nichtkommutative Galoissche Theorie und die Relation* $\xi_{\lambda,\mu,\nu}\xi_{\lambda,\mu\nu,\pi}\xi_{\mu,\nu,\pi}^{\lambda} = \xi_{\lambda,\mu,\nu\pi}\xi_{\lambda,\mu,\nu,\pi}$, Deutsche Math. **5** (1940), 138–149.
4. R. Brauer, *Über Systeme hypercomplexer Zahlen*, Math. Z. **30** (1929), 79–107.
5. Nathan Jacobson, *Construction of central simple associative algebras*, Ann. of Math. (2) **45** (1944), 653–666.
6. _____, *The theory of rings*, Amer. Math. Soc., Providence, RI, 1943.
7. Martin Lorenz, *Division algebras generated by finitely generated nilpotent groups*, J. Algebra **85** (1983), no. 2, 368–381.
8. E. Noether, *Nichtkommutative Algebra*, Math. Z. **37** (1933), 514–541.
9. D. S. Passman, *Semiprime and prime crossed products*, J. Algebra **83** (1983), no. 1, 158–178.
10. V. P. Platonov and V. I. Yanchevskii, *Dieudonne's conjecture on the structure of unitary groups over a skew-field and Hermitian K-theory*, Izv. Akad. Nauk SSSR Ser. Mat. **48** (1984), no. 6, 1266–1294; English transl. in Math. USSR-Izv. **25** (1985).
11. _____, *On the theory of Henselian division algebras*, Dokl. Akad. Nauk SSSR **297** (1987), 294–298; English transl. in Soviet Math. Dokl. **36** (1988).
12. _____, *Finite-dimensional Henselian division algebras*, Dokl. Akad. Nauk SSSR **297** (1987), 542–547; English transl. in Soviet Math. Dokl. **36** (1988).
13. Jean-Pierre Tignol, *On the corestriction of central simple algebras*, Math. Z. **194** (1987), no. 2, 267–274.

14. A. I. Tikhomirov, *Eine Verallgemeinerung des Begriffes des verschrankten Produktes*, Izv. Akad. Nauk SSSR Ser. Mat. **5** (1941), 297–304. (Russian)
15. Jean-Pierre Tignol, *Generalized crossed products*, Preprint, Inst. Math. Pure Appl., Louvain, Belgium, 1987.

Institute of Mathematics
Academy of Sciences of the Belorussian SSR

Received 7/DEC/87

Translated by H. H. McFADEN

On the Metaplectic Kernel for Anisotropic Groups

UDC 512.743

A. S. RAPINCHUK

Abstract. Under certain assumptions the metaplectic kernel $M(G, S)$ is calculated for the group $G = SL(1, D)$, where D is a quaternion algebra over an algebraic number field K.

Let G be an almost simple simply connected algebraic group defined over a field K of algebraic numbers, and let S be some finite subset of the set V^K of all nonequivalent valuations of the field K. The metaplectic kernel $M(G, S)$ of G with respect to the set S is defined to be the kernel of the canonical restriction mapping

$$H^2(G(A(S))) \to H^2(G(K)),$$

where $G(A(S))$ is the S-adèle group of G, $G(K)$ is the group of K-rational points, and $H^2(*)$ denotes the second continuous cohomology group with coefficients in the one-dimensional torus $\mathbf{R}/\mathbf{Z}$ (here the topology on $G(K)$ is assumed to be discrete). It is well known that the computation of $M(G, S)$ is one of the crucial points in the solution of the congruence problem for S-arithmetic pogroups G (see [1], [2]). The investigation of $M(G, S)$ for K-isotropic groups, which began with the fundamental work of Moore [3], has been practically completed in the work of Prasad and Raghunathan ([4], [5]) and of Bak and Rehmann ([6], [7]). Nevertheless, $M(G, S)$ has so far not been computed for K-anisotropic groups. In the present article we analyze the minimal case of three-dimensional (anisotropic) groups. The assertions obtained in it, together with results in [8]–[10], enable us to investigate the central "part" of the congruence kernel in the given situation.

It is well known that a three-dimensional simply connected K-anisotropic group G is an algebraic group associated with the group $SL(1, D)$ of

1991 *Mathematics Subject Classification*. Primary 20G35.
Translation of Dokl. Akad. Nauk BSSR **29** (1985), no. 12, 1068–1071.

elements with reduced norm equal to the identity of a certain division algebra D of quaternions over K. Let $S \subset V^K$ be a finite subset containing all Archimedean valuations and such that $D_v = D \otimes_K K_v \simeq M_2(K_v)$ for all $v \notin S$.

THEOREM. *If S contains a non-Archimedean valuation v_1 such that $D_{v_1} \simeq M_2(K_{v_1})$, then the metaplectic kernel $M(G, S)$ has exponent ≤ 2. In the general case $M(G, S)$ is isomorphic to a subgroup of the extension of a group of exponential 2 by the group $E(K)$ of all roots of unity contained in K.*

The proof of the theorem is based on reduction of the computation of $M(G, S)$ to the investigation of reciprocity laws on certain tori. The fundamental difference from the earlier variants of this method is that in treating anisotropic groups it is necessary to work with anisotropic tori.

Accordingly, let $T \subset G$ be a K-defined torus. Consider the commutative diagram

$$\begin{array}{ccc} H^2(G(A(S))) & \xrightarrow{\varphi} & H^2(G(K)) \\ \rho \downarrow & & \downarrow \psi \\ H^2(T(A(S))) & \xrightarrow{\theta} & H^2(T(K)) \end{array} \quad (1)$$

in which all the arrows are restriction homomorphisms. It is known (see [11]) that $H^2(G(A(S))) = \prod_{v \notin S} H^2(G(K_v))$, so that an element $x \in H^2(G(A(S)))$ can be written in the form $x = (x_v)$, where $x_v \in H^2(G(K_v))$. The proof that $M(G, S)^2 = 1$ is carried out by the following scheme. Let $y = x^2 \in M(G, S)^2$, $y \neq 1$. Then $x = (x_v)$, $y = (y_v)$, and for some $v_0 \in V^K \backslash S$ the component y_{v_0} is not $= 1$. We now use the following assertion, due to Moore [3].

LEMMA 1. *Let v be a non-Archimedean valuation and let T be a K_v-decomposable torus in G. Then the group $H^2(G(K_v))$ is a cyclic group of order $\mu_v = [E(K_v)]$ and is generated by a cocycle $\alpha_v(s, t)$ such that $\alpha_v(s, t) = (\chi(s), \chi(t))_v$ for all $s, t \in T(K_v)$, where $\chi: T \to \mathbf{G}_m$ is some K_v-isomorphism, and $(*, *)_v$ is the norm residue symbol of degree μ_v (see [12]).*

Subjecting the choice of T to certain conditions, in particular, requiring that T be decomposable over K_{v_0}, we get that

$$\rho_v(x_v) = \begin{cases} \alpha_v^{n_v} & \text{if } T \text{ is } K_v\text{-decomposable}, \\ 1 & \text{if } T \text{ is } K_v\text{-anisotropic}, \end{cases} \quad (2)$$

where the n_v are natural numbers, and $2n_{v_0}$ is not divisible by μ_{v_0}. Since $\varphi(x) = 1$, it follows that $\theta(\rho(x)) = 1$, which leads to a relation ("reciprocity law") of the form

$$\prod (\alpha_v(s, t))^{2n_v} = 1 \quad \forall s, t \in T(K).$$

It remains to see that such relations do not exist. Here we use class field theory and the following description of the cocycle α_v.

It is easy to see that T is isomorphic to an algebraic group determined by the group $SL(1, L|K) = \{z \in L | N_{L|K}(z) = 1\}$, where L is some quadratic extension of K contained in D. Let σ be a generator of the group $\text{Gal}(L|K)$. Then by Hilbert's Theorem 90, $SL(1, L|K) = \{\frac{\sigma a}{a} | a \in L^*\}$. We have

LEMMA 2. *If* $L \subset K_v$, *then*

$$\alpha_v\left(\frac{\sigma a}{a}, \frac{\sigma b}{b}\right) = \left(\frac{a}{\sigma a}, b\right)_{w'} \left(\frac{a}{\sigma a}, b\right)_{w''},$$

where w' and w'' are two extensions of the valuation v to L.

(Here $(*, *)_{w'}$ is the symbol of the norm residue on $L_{w'}^* = K_v^*$ of degree $\mu_v = [E(K_v)] = [E(L_{w'})]$.) We also need

LEMMA 3. *There exists a finite subset $S_0 \subset V^K$ such that for any $v \in V^K \setminus S_0$ and any K_v-anisotropic torus $T \subset G$ the restriction mapping*

$$H^2(G(K_v)) \to H^2(T(K_v))$$

is trivial.

PROOF OF THE THEOREM. It is known (see [5]) that $M(G, S)$ is a finite group. We assume that the subgroup $M(G, S)^2$ of squares is not $= 1$, and let p be a prime divisor of its order and $y = x^2$ an element of order p. Then $x = (x_v)$, $y = (y_v)$, and for some $v_0 \in V^K \setminus S$ the component y_{v_0} is not $= 1$. Let $T \subset G$ be a K-torus that is decomposable over completions with respect to valuations in $(S_0 \setminus S) \cup \{v_0, v_1\}$ (here S_0 is exclusively the subset in Lemma 3). The existence of such a torus follows, for example, from the weak approximation property for the variety $\mathcal{T}$ of tori of the group G, which in turn is a consequence of its K-rationality (see [13]). We consider the corresponding diagram (1) and the representation $\rho(x) = (\rho_v(x_v))$ of the form (2). Then the bilinear mapping $\delta(s, t) = \prod_{v \in S_1} \alpha_v(s, t)^{n_v}$, where $S_1 = \{v \in V^K \setminus S | T$ is K_v-decomposable$\}$ determines a trivial cocycle in $H^2(T(K))$. This implies that $\delta(s, t)$ is symmetric, and hence $\delta(s, t)^2 = 1$ in view of the skew-symmetry of the norm residue symbol, i.e., $\prod_{v \in S_1} \alpha_v(s, t)^{2n_v} = 1$ for all $s, t \in T(K)$. Since $\rho(y_v) = \alpha_v^{2n_v}$ and y is an element of order p, for all $v \in S_1$ the greatest common divisor $(\mu_v, 2n_v)$ is equal either to μ_v or to μ_v/p, and $(\mu_{v_0}, 2n_v) = \mu_{v_0}/p$. Therefore, the components $\rho_v(x_v)$ different from 1 have the form $\rho_v(x_v) = \beta_v^{m_v}$, where the function β_v is constructed by starting from the symbol $[*, *]_v$ of the norm residue on K_v^* of degree (μ_v, p). Thus, considering Lemma 2, we get that

$$\prod_{v \in S_1} \left(\prod_{w | v} \left[\frac{a}{\sigma a}, b\right]_w\right)^{m_v} = 1 \quad \forall a, b \in L^*, \qquad (3)$$

where w is an extension of v to L (recall that $T \simeq SL(1, L|K))$. We show that (3) is impossible if $(m_{v_0}, p) = 1$.

Consider the field $M = L(\zeta_p)$, where ζ_p is a primitive pth root of unity, and denote by $\{*, *\}_u$ the symbol of the norm residue of degree p with respect to the valuation u of the field M. Using the weak approximation theorem, we choose an element $a \in L^*$ such that the element $a/\sigma a$ is uniformizing with respect to $w'_0 | v_0$ and $w'_1 | v_1$ (the notation corresponds to Lemma 2). Suppose also that u_0 and u_1 are extensions of w'_0 and w'_1 to M, and p_1 is the corresponding prime number. In this case if $p \neq p_1$, then the extension $M_{u_1} | L_{w'_1}$ is unramified; however, if $p = p_1$, then it is totally ramified, but the ramification index satisfies $e = [M_{u_1} : L_{w'_1}] \leq p - 1$. In both cases the value $u_1(a/\sigma a)$ of the valuation is relatively prime to p, so that there exists a $c_{u_1} \in M^*_{u_1}$ with the property $\{a/\sigma a, c_{u_1}\}_{u_1} = \zeta_p^{-1}$. Similarly, since $M_{u_0} = L_{w'_0}$, it follows that there exists a $c_{u_0} \in M^*_{u_0}$ with the property $\{a/\sigma a, c_{u_0}\}_{u_0} = \zeta_p$. We find ourselves under the conditions of applicability of the following assertion (see [12]).

LEMMA 4. *Let M be a field of algebraic numbers that contains a primitive nth root of unity, let $(*, *)_u$ be the symbol of the norm residue of degree n on M^*_u and let $a \in M^*$. Assume that for each valuation u of the field M an nth root θ_u of unity is given, and the following conditions hold:*
(i) $\theta_u = 1$ *for almost all u;*
(ii) $\prod_u \theta_u = 1$;
(iii) *for any u there exists a $c_u \in M^*_u$ such that $(a, c_u) = \theta_u$.*
Then there exists a $c \in M^*$ such that $(a, c)_u = \theta_u$ for all u.

Accordingly, there exists a $c \in M^*$ with the property

$$\left\{\frac{a}{\sigma a}, c\right\}_u = \begin{cases} \zeta_p, & u = u_0, \\ \zeta_p^{-1}, & u = u_1, \\ 1, & u \neq u_0, u_1. \end{cases}$$

We show that (3) does not hold for a and $b = N_{M|L}(c)$. To do this it suffices to establish that

$$\left[\frac{a}{\sigma a}, b\right]_w = \begin{cases} \zeta_p, & w = w'_0, \\ 1 & \text{if } w \neq w'_0 \text{ and } w|v, \text{ where } v \in V^K \setminus S. \end{cases} \quad (4)$$

The equality (4) is obvious if μ_w is not divisible by p. Otherwise, L_w contains a primitive pth root of unity, and

$$\left[\frac{a}{\sigma a}, b\right]_w = \prod_{u|w} \left\{\frac{a}{\sigma a}, c\right\}_u = \begin{cases} \zeta_p, & w = w'_0, \\ 1, & w \neq w'_0 \text{ and } w|v, \text{ where } v \in V^K \setminus S. \end{cases}$$

The first part of our theorem is proved.

The proof of the second part is obtained by using the first density theorem of Chebotarev (see [12]) and the following simple assertion.

LEMMA 5. *Let $S_1 \supset S_2$. Then there exists a natural homomorphism* $\pi = \pi(S_1, S_2): M(G, S_1) \to M(G, S_2)$ *whose cokernel is contained in* $\prod_{v \in S_1 \setminus S_2} H^2(G(K_v))$.

Generalizations of the results obtained to other anisotropic groups, as well as an application of them to the computation of a congruence kernel, will be given in subsequent publications of the author.

The author takes it as his pleasant duty to express deep gratitude to V. P. Platonov for attention to this work and for valuable advice.

Bibliography

1. H. Bass, J. Milnor, and J.-P. Serre, *Solution of the congruence subgroup problem for SL_n ($n \geq 3$) and Sp_{2n} ($n \geq 2$)*, Inst. Hautes Études Sci. Publ. Math. **33** (1967), 59–137.
2. V. P. Platonov, *Arithmetic theory of algebraic groups*, Uspekhi Mat. Nauk **37** (1982), no. 3, 3–54; English transl. in Russian Math. Surveys **37** (1982).
3. Calvin C. Moore, *Group extensions of p-adic and adelic linear groups*, Inst. Hautes Études Sci. Publ. Math. **35** (1968), 157–222.
4. Gopal Prasad and M. S. Raghunathan, *Topological central extensions of semisimple groups over local fields*, Ann. of Math. (2) **119** (1984), no. 1, 143–201.
5. ____, *On the congruence subgroup problem: determination of the "metaplectic kernel"*, Invent. Math. **71** (1983), no. 1, 21–42.
6. Anthony Bak and Ulf Rehmann, *The congruence subgroup and metaplectic problems for $SL_{n \geq 2}$ of division algebras*, J. Algebra **78** (1982), no. 2, 475–547.
7. Anthony Bak, *Le probleme des sous-groupes de congruence et le probleme metaplectique pour les groupes classiques de rang > 1*, C. R. Acad. Sci. Paris Sér. I Math. **292** (1981), no. 5, 307–310.
8. V. P. Platonov and A. S. Rapinchuk, *On the group of rational points of three-dimensional groups*, Dokl. Akad. Nauk SSSR **247** (1979), 279–282; English transl. in Soviet Math. Dokl. **20** (1979).
9. V. P. Platonov and A. S. Rapinchuk, *Multiplicative structure of division algebras over number fields and the Hasse norm principle*, Trudy Mat. Inst. Steklov. **165** (1984), 171–187; English transl. in Proc. Steklov Inst. Math. **165** (1985).
10. G. A. Margulis, *Multiplicative groups of a quaternion algebra over a global field*, Dokl. Akad. Nauk SSSR **252** (1980), 542–545; English transl. in Soviet Math. Dokl. **21** (1980).
11. M. S. Raghunathan, *On the congruence subgroup problem*, Inst. Hautes Études Sci. Publ. Math. **46** (1976), 107–161.
12. J. W. S. Cassels and A. Frölich (eds.), *Algebraic number theory*, Academic Press, London, and Thompson, Washington DC, 1967.
13. C. Chevalley, *On algebraic group varieties*, J. Math. Soc. Japan **6** (1954), no. 3-4, 303–324.

Institute of Mathematics
Academy of Sciences of the Belorussian SSR

Received 14/DEC/84

Translated by H. H. McFADEN

The Metaplectic Kernel for the Group $SL(1, D)$

A. S. RAPINCHUK

> Abstract. The metaplectic kernel is calculated for the algebraic group G over an algebraic number field K, defined by the group $SL(1, D)$ which consists of all elements of the finite-dimensional central skew field D with reduced norm 1.

Let G be a simply connected almost simple algebraic group defined over a field K of algebraic numbers. For any finite subset S of the set V^K of all valuations of K, with S containing all the Archimedean valuations, denote by $A(S)$ the ring of S-adèles of the field K. If B is some K-algebra, then denote by $G(B)$ the group of B-points of the group G; in particular, $G(A(S))$, $G(K)$, and $G(K_v) = G_v$ are the respective groups of $A(S)$-points, K-points, and points over the v-adic completion K_v of K. The *metaplectic kernel* $M(G, S)$ of G with respect to the set S is defined to be the kernel of the restriction mapping $H^2(G(A(S))) \to H^2(G(K))$, where $H^2(*)$ denotes the second continuous cohomology group with coefficients in the one-dimensional torus $\mathbf{R}/\mathbf{Z}$ (the topology on $G(K)$ is discrete).

The problem of determining the metaplectic kernel $M(G, S)$ is closely connected with the solution of the congruence problem (see [1], [2]), which has stimulated numerous investigations in this area. The case of isotropic groups has by now been almost exhaustively treated (Prasad and Raghunathan [3], [4]; see also Bak and Rehmann [5], [6]). Nevertheless, there are so far no investigations of $M(G, S)$ for anisotropic groups. The present note is devoted to the computation of $M(G, S)$ for groups G defined by $SL(1, D)$, where D is a finite-dimensional central division algebra over the field K (recall that $SL(1, D)$ is the collection of elements in D^* with reduced norm equal to 1). It is well known that this class of groups, as a rule, adequately reflects the general situation (see, for example, [7]–[9]). The

---

1991 *Mathematics Subject Classification.* Primary 12E15, 20G35.
Translation of Dokl. Akad. Nauk BSSR **30** (1986), no. 3, 197–200.

problem of determining the metaplectic kernel is not exceptional in this respect: to a significant degree it reduces to $SL(1, D)$ for the classical groups (cf. [6]).

We proceed to a formulation of the main results. Let D be a central division algebra of index n over a field K of algebraic numbers, and let G be the algebraic K-group associated with the group $SL(1, D) = \{x \in D^* | \mathrm{Nrd}_{D|K}(x) = 1\}$, where $\mathrm{Nrd}_{D|K}: D^* \to K^*$ is the reduced norm homomorphism. We also fix a finite subset $S \subset V^K$ containing the set V_∞^K of Archimedean valuations. The exceptional subset corresponding to D and S is defined to be the set $S_e = \{v \in V^K \setminus S | D_v = D \otimes_K K_v \simeq M_2(F_v)\}$, where F_v is a division algebra over the field K_v. Note that the cardinality s_e of the set S_e is finite if $n > 2$, and is equal to ω_0 if $n = 2$.

BASIC THEOREM. *If S has a non-Archimedean valuation v_0 such that $D_{v_0} \simeq M_n(K_{v_0})$, then the metaplectic kernel $M(G, S)$ is a finite subgroup of the group $B(D, S) = (\mathbf{Z}/2\mathbf{Z})^{s_e}$. In the general case $M(G, S)$ is isomorphic to a finite subgroup of the product $B(D, S) \times E(K)$, where $E(K)$ is the group of all roots of unity contained in K.*

COROLLARY. *If $S_e = \emptyset$, in particular, if the index of the division algebra D is odd, then*

$$M(G, S) = \begin{cases} 1 & \text{if } D_{v_0} \simeq M_n(K_{v_0}); \text{ for some } v_0 \in S \\ & \text{that is not totally imaginary}; \\ \text{a subgroup of } E(K) & \text{otherwise.} \end{cases}$$

We remark that the case of a division algebra of quaternions (i.e., when $n = 2$) was considered earlier in [16], therefore, it can be assumed in what follows that $n > 2$. As in [16], the key role in the proof of the basic theorem is played by

PROPOSITION 1. *Assume that S contains a non-Archimedean valuation v_0 such that $D_{v_0} \simeq M_n(K_{v_0})$. Then $M(G, S)$ is isomorphic to a subgroup of the group $B(D, S)$.*

Below we present the scheme of proof for Theorem 1. We need the following additional notation. Let $S_0 = \{v \in V^K \setminus S | D_v \text{ is a division algebra}\}$, and let $S_1 = \{v \in V^K \setminus S | D_v \neq M_n(K_v)\}$. Further, denote by G_{S_1} the product $\prod_{v \in S_1} G_v$, and by δ the diagonal imbedding of $G(K)$ in G_{S_1}. Proposition 1 is a consequence of the following assertions.

PROPOSITION 2. *Let U be an open subgroup of G_{S_1}. Then the kernel of the restriction mapping $H^2(G(A(S \cup S_1))) \to H^2(\delta^{-1}(U))$ does not contain nontrivial elements of finite order.*

PROPOSITION 3. *Let $N(G, S_1) = \mathrm{Ker}(H^2(G_{S_1}) \overset{\mathrm{res}}{\to} H^2(G(K)))$. Then there exists a unique imbedding of $N(G, S_1)$ in $B(D, S)$.*

We assume for the time being that we have proved Propositions 2 and 3, and we deduce Proposition 1. It follows from the Künneth formula (see [10]) that $H^2(G(A(S))) = H^2(G_{S_1}) \times H^2(G(A(S \cup S_1)))$. Denote by p_1 and p_2 the projections of $H^2(G(A(S)))$ on the first and second factors, respectively. Now let $x = (p_1(x), p_2(x)) \in M(G, S_1)$. Since x is continuous, there exists an open subgroup $U \subset G_{S_1}$ such that $p_1(x)$ lies in the kernel of the restriction mapping $H^2(G_{S_1}) \to H^2(U)$. Then $p_2(x)$ lies in the kernel of the mapping $H^2(G(A(S \cup S_1))) \to H^2(\delta^{-1}(U))$. Therefore, $p_2(x) = 1$ in view of Proposition 2 and the finiteness of $M(G, S)$ (see [4], [10]). But then $p_1(x) \in N(G, S_1)$, so that the projection p_1 actually induces an imbedding of $m(G, S)$ in $N(G, S_1)$. To conclude the proof of Proposition 1 it now suffices to invoke Proposition 3.

The proof of Proposition 2 is based on reduction to the investigation of the reciprocity laws on certain tori. The fundamental difference from the previously existing variants of this method (which go back to Moore [11] and Matsumoto [12]) is that in the consideration of anisotropic groups it is necessary to work with anisotropic tori.

LEMMA 1 (Matsumoto [12]). *Let* $v \in V^K \setminus V_\infty^K$. *Then* $H^2(SL_n(K_p))$ $(n \geq 2)$ *is a cyclic group of order* $\mu_v = [E(K_v)]$ *and is generated by the class of a cocycle* $\alpha_v(a, b)$ *such that*

$$\alpha_v(\mathrm{diag}(s_1, \ldots, s_n), \mathrm{diag}(t_1, \ldots, t_n)) = \prod_{n \geq i \geq j} (s_i, t_j)_v$$

for all $s_i, t_j \in K_v^*$ *with* $s_1 s_2 \cdots s_n = t_1 t_2 \cdots t_n = 1$, *where* $(\ ,\ )_v$ *is the symbol for the norm residue on* K_v^* *of degree* μ_v.

In reality we need a result describing the restriction of α_v to a broader class of tori.

LEMMA 2. *Let* L_w *be an extension of* K_v *of degree* $m|n$. *Let* $d = n/m$ *and denote by* T *the torus* $(L_w^* \times \cdots \times L_w^*) \cap SL_n(K_v)$. *Then the range of the restriction mapping* $H^2(SL_n(K_v)) \to H^2(T)$ *is contained in the subgroup generated by the class of a cocycle* θ_v *such that*

$$\theta_v(\mathrm{diag}(a_1, \ldots, a_d), \mathrm{diag}(b_1, \ldots, b_d)) = \prod_{d \geq i \geq j} (a_i, b_j)_w,$$

where $(\ ,\ )_w$ *is the symbol for the norm residue on* L_w^* *of degree* $\mu_v = [E(K_v)]$.

It follows from the Grunwald-Wang theorem (see [13]) that D always has maximal subfields L that are cyclic extensions of the center (of degree n). Further, by Hilbert's Theorem 90, the group $SL(1, L|K) = \{x \in L^* | N_{L|K}(x) = 1\}$ coincides with the set $\{\sigma a/a | a \in L^*\}$, where σ is a generator of the Galois group $\mathrm{Gal}(L|K)$. Denote by T the algebraic K-torus determined by the group $SL(1, L|K)$. Using Lemma 2, we can compute the restriction of the cocycle α_v to $T(K)$.

LEMMA 3. *Let $d_v = n/m_v$, where $m_v = [LK_v : K_v]$, and let $w_1, \ldots, w_d$ be distinct extensions of the valuation v to L. Then the range of the restriction mapping $\rho_v : H^2(SL_n(K_v)) \to H^2(T(K_v))$ is contained in the subgroup generated by the class of a cocycle δ_v such that*

$$\delta_v\left(\frac{\sigma a}{a}, \frac{\sigma b}{b}\right) = \left(\frac{a}{\sigma^{d_v} a}, b\right)_{w_1} \prod_{i=1}^{d_v} \left(\frac{a}{\sigma^{-1} a}, b\right)_{w_i} \quad \forall a, b \in L^*,$$

and coincides with it when $m_v = 1$.

Suppose now that $x \in \text{Ker}(H^2(G(A(S \cup S_1))) \to H^2(\delta^{-1}(U)))$ has finite order. It is known (see [10]) that $H^2(G(A(S \cup S_1))) = \prod_{v \notin S \cup S_1} H^2(G_v)$, hence $x = (x_v)$, where $x_v \in H^2(G_v)$. By Lemma 1, $x_v = \alpha_v^{i_v}$, and we show that the numbers i_v are multiples of μ_v for all $v \notin S \cup S_1$. Suppose that this does not hold for some $v_1 \notin S \cup S_1$. We choose a maximal cyclic subfield $L \subset D$ in such a way that the local degree $m_{v_1} = [LK_{v_1} : K_{v_1}]$ is equal to 1, and we denote by T the algebraic K-torus associated with the group $SL(1, L|K)$. We have the commutative diagram

$$\begin{array}{ccc} H^2(G(A(S \cup S_1))) & \xrightarrow{\varphi} & H^2(\delta^{-1}(U)) \\ {\scriptstyle \rho} \downarrow & & \downarrow {\scriptstyle \psi} \\ H^2(T(A(S \cup S_1))) & \xrightarrow{\kappa} & H^2(T(K) \cap \delta^{-1}(U)) \end{array}$$

in which all arrows are restriction mappings. Since $\varphi(x) = 1$, it follows that $\kappa(\rho(x)) = \psi(\varphi(x)) = 1$, therefore, $\rho(x) \in \text{Ker}\,\kappa$. On the other hand, $\rho_v(\alpha_v) = \delta_v^{j/v}$ by Lemma 3, where j_{v_1} is relatively prime to μ_{v_1}. Therefore, $\rho_v(x_v) = \delta_v^{k_v}$, where k_{v_1} is not divisible by μ_{v_1}. Bearing in mind that the cocycle $\rho(x)$ on $T(A(S \cup S_1))$ is induced by the element $(\rho_v(x_v))$, we see easily that the condition $\rho(x) \in \text{Ker}\,\kappa$ means the following: the cocycle δ on $T(K) \cap \delta^{-1}(U)$ defined by the formula

$$\delta(a, b) = \prod_{v \notin S \cup S_1} \delta_v(a, b)$$

is trivial; in particular, $\delta(a, b) = \delta(b, a)$ for all $a, b \in T(K) \cap \delta^{-1}(U)$. By using the description of δ_v in Lemma 3 it is not hard to get from this that

$$\prod_{v \notin S \cup S_1} \left(\left(\frac{a^2}{\sigma^{d_v} a \sigma^{-d_v} a}, b\right)_{w_1} \prod_{i=1}^{d_v} \left(\frac{a^2}{\sigma a \sigma^{-1} a}, b\right)_{w_i} \right)^{k_v} = 1 \quad (1)$$

for all $a, b \in L^*$ such that $\sigma a/a$, $\sigma b/b \in U$ (where $w_1, \ldots, w_{d_v}$ are distinct extensions of v to L). The relation (1) connecting the symbols of the norm residues is in a certain sense a reciprocity law, so that our problem has actually been reduced to the description of those values k_v for which a reciprocity law of the form (2) really does hold. It is known from class field theory (see [14]) that a reciprocity law of the form $\prod_v (a, b)_v^{\mu_v/\mu} = 1$ is valid in any field of algebraic numbers (the product is over *all* valuations, and any other reciprocity law is a power of it (see [11])). It turns out that for relations of the form (2) there is an analogous result, from which in view of $k_{v_0} = 0$ it follows that k_{v_1} is divisible by μ_{v_1}. This contradiction proves Proposition 2.

The proof of Proposition 3 goes as follows. It is first shown that the restriction to $N(G, S_1)$ of the projection $q_2: H^2(G_{S_1}) \to H^2(G_{S_2})$, where $S_2 = S_1 \setminus S_0$, is injective (note that $H^2(G_{S_1}) = H^2(G_{S_0}) \times H^2(G_{S_2})$). It is easy to see that this fact is equivalent to the following assertion, whose proof uses recent results of Raghunathan generalizing [15].

PROPOSITION 4. *The restriction mapping $H^2(G_{S_0}) \to H^2(G(K))$ is injective.*

Continuing the arguments, we consider the projection $q_0: H^2(G_{S_1}) \to H^2(G_{S_0})$. If $x = (q_0(x), q_2(x)) \in N(G, S_1)$, then there exists an open subgroup $U \subset G_{S_0}$ such that $q_0(x)$ lies in the kernel of the restriction mapping $H^2(G_{S_0}) \to H^2(U)$. Then $q_2(x) \in N(G, S_2, U) = \mathrm{Ker}(H^2(G_{S_2}) \to H^2(G(K) \cap U))$. Thus,

$$N(G, S_1) \simeq q_2(N(G, S_1)) \subset \bigcup_U N(G, S_2, U)$$

(the union is over all open subgroups $U \subset G_{S_2}$), and

$$N(G, S_2, U_1) \cup N(G, S_2, U_2) \subset N(G, S_2, U_1 \cap U_2).$$

Therefore, everything reduces to the proof of the following assertion.

PROPOSITION 5. *For an arbitrary open subgroup $U \subset G_{S_2}$ there exists an imbedding of the group $N(G, S_2, U)$ in $B(D, S)$.*

The author expresses deep gratitude to V. P. Platonov for advice and for a discussion of this work.

BIBLIOGRAPHY

1. H. Bass, J. Milnor, and J.-P. Serre, *Solution of the congruence subgroup problem for $SL_n(n \geq 3)$ and $Sp_{2n}(n \geq 2)$*, Inst. Hautes Études Sci. Publ. Math. **33** (1967), 59–137.
2. V. P. Platonov, *Arithmetic theory of algebraic groups*, Uspekhi Mat. Nauk **37** (1982), no. 3, 3–54; English transl. in Russian Math. Surveys **37** (1982).
3. Gopal Prasad and M. S. Raghunathan, *Topological central extensions of semisimple groups over local fields*, Ann. of Math. (2) **119** (1984), no. 1, 143–201.
4. _____, *On the congruence subgroup problem: determination of the "metaplectic kernel"*, Invent. Math. **71** (1983), no. 1, 21–42.

5. Anthony Bak and Ulf Rehmann, *The congruence subgroup and metaplectic problems for $SL_{n\geq 2}$ of division algebras*, J. Algebra **78** (1982), no. 2, 475–547.
6. Anthony Bak, *Le probleme des sous-groupes de congruence et le probleme metaplectique pour les groupes classiques de rang > 1*, C. R. Acad. Sci. Paris Sér. I Math. **292** (1981), no. 5, 307–310.
7. V. P. Platonov, *The problem of strong approximation and the Kneser-Tits hypothesis for algebraic groups*, Izv. Akad. Nauk SSSR Ser. Mat. **33** (1969), 1211–1219; English transl. in Math. USSR-Izv. **3** (1969).
8. _____, *Algebraic groups and reduced K-theory*, Proc. Internat. Congr. Math. (Helsinki, 1978), Acad. Sci. Fennica, Helsinki, 1980, pp. 311–317.
9. Jacques Tits, *Groupes de Whitehead de groupes algebriques simples sur un corps (d'apres V. P. Platonov et al.)*, Sèminaire Bourbaki, 29e annee (1976/77), Exp. No. 505, Lecture Notes in Math., vol. 677, Springer-Verlag, Berlin and New York, 1977, pp. 218–236.
10. M. S. Raghunathan, *On the congruence subgroup problem*, Inst. Hautes Études Sci. Publ. Math. **46** (1976), 107–161.
11. Calvin C. Moore, *Group extensions of p-adic and adelic linear groups*, Inst. Hautes Études Sci. Publ. Math. **35** (1968), 157–222.
12. Hideya Matsumoto, *Sur les sous-groupes arithmetiques des groupes semi-simples deployés*, Ann. Sci. École Norm. Sup. (4) **2** (1969), 1–62.
13. Shianghaw Wang, *On Grunwald's theorem*, Ann. of Math. (2) **51** (1950), no. 2, 471–484.
14. J. W. S. Cassels and A. Frölich (eds.), *Algebraic number theory*, Academic Press, London, and Thompson, Washington DC, 1967.
15. V. P. Platonov and A. S. Rapinchuk, *Multiplicative structure of division algebras over number fields and the Hasse norm principle*, Trudy Mat. Inst. Steklov. **165** (1984), 171–187; English transl. in Proc. Steklov Inst. Math. **165** (1985).
16. A. S. Rapinchuk, *The metaplectic kernel for anisotropic groups*, Dokl. Akad. Nauk BSSR **29** (1985), no. 12, 1068–1071. (Russian)

Institute of Mathematics
Academy of Sciences of the Belorussian SSR

Received 15/MAR/85

Translated by H. H. McFADEN

On Finite Presentability of Reduced Norms in Simple Algebras

UDC 512.7+511

A. S. RAPINCHUK

Abstract. Let D be a finite-dimensional central skew field over a field K, and let $\operatorname{Nrd}_{D/K}: D^* \to K^*$ be the reduced norm homomorphism. The question when can $\operatorname{Nrd}_{D/K}(D^*)$ be generated by a finite number of norm subgroups $N_{F/K}(F^*)$ attached to maximal subfields $F \subset D$ is investigated.

Let D be a finite-dimensional central division algebra of index n over a field K, and let $\operatorname{Nrd}_{D/K}: D^* \to K^*$ be the reduced norm homomorphism (see [4]). If $x \in D$, then $\operatorname{Nrd}_{D/K}(x) = N_{F/K}(x)$ is the usual norm in the field extension, where $F \subset D$ is the maximal subfield containing x. Therefore,

$$\operatorname{Nrd}_{D/K}(D^*) = \bigcup_F N(F/K),$$

where $N(F/K) = N_{F/K}(F^*)$ is the norm subgroup, and the union is over all maximal subfields of D. Since the problem of describing all maximal subfields and the corresponding norm subgroups can turn out not to be simple, it is natural to ask when $\operatorname{Nrd}_{D/K}(D^*)$ is presentable by a finite collection of norm subgroups. In particular, when is $\operatorname{Nrd}_{D/K}(D^*)$ generated by $\bigcup_{i=1}^r N(F_i/K)$ for some finite collection $F_1, \ldots, F_r$ of maximal subfields of the division algebra D?

This question is easy to answer in the positive for a locally compact field K. Next in complexity is the case of number fields. Here even a division algebra of quaternions presents surprises.

PROPOSITION 1. *Let D be a division algebra of quaternions over a field K of algebraic numbers. Then $\operatorname{Nrd}_{D/K}(D^*)$ is generated by finitely many norm subgroups corresponding to maximal subfields of D if and only if D splits over*

1991 *Mathematics Subject Classification.* Primary 12E15; Secondary 11R52, 16E20.
Translation of Dokl. Akad. Nauk BSSR **32** (1988), no. 1, 5–8.

all Archimedean valuations v of the field K, i.e., $D_v = D \otimes_K \simeq M_2(K_v)$. In particular, the division algebra $H = (\frac{-1,-1}{\mathbb{Q}})$ of Hamiltonian quaternions is not finitely generated.

PROOF. Denote by V^K the set of all nonequivalent valuations of the field K and by V_∞^K the subset of Archimedean valuations, and let $S_0 = \{v \in V^K | D_v$ is a division algebra$\}$. If $S_0 \subset V^K \setminus V_\infty^K$, then $[K_v^* / K_v^{*2}] \geq 4$ for any $v \in S_0$, and hence there exist $a_v, b_v \in K_v^* \setminus K_v^{*2}$ such that $a_v b_v \notin K_v^{*2}$. By the weak approximation theorem, there exist $a, b \in K^*$ with the property $a \in a_v K_v^{*2}$ and $b \in b_v K_v^{*2}$ for all $v \in S_0$. Then the fields $F_1 = K(\sqrt{a})$, $F_2 = K(\sqrt{b})$, and $F_3 = K(\sqrt{ab})$ are isomorphically imbedded in D. On the other hand, the index in K^* of the subgroup $N(F_1/K)N(F_2/K)N(F_3/K)$ does not exceed 2 (see [6], Exercise 5.2), which implies that $\mathrm{Nrd}_{D/K}(D^*)$ is generated by norm subgroups corresponding to at most four maximal subfields.

Conversely, let $L_1, \ldots, L_r \subset D$ be maximal subfields such that $\bigcup_{i=1}^r N(L_i/K)$ generates $\mathrm{Nrd}_{D/K}(D^*)$. Assume that there exists an Archimedean valuation v of K such that the algebra D_v remains a division algebra. In this case if $L_i = K(\sqrt{d_i})$, then $d_i < 0$ in K_v for all $i = 1, \ldots, r$. We consider the composite

$$L = L_1 \cdots L_r = K(\sqrt{d_1}, \ldots, \sqrt{d_r}).$$

Suppose that the images of $d_1, \ldots, d_l$ are multiplicatively independent in the factor group K^*/K^{*2} and generate the whole group $\langle d_1, \ldots, d_r \rangle K^{*2}$. Then there exists an automorphism $\sigma \in \mathrm{Gal}(L/K)$ such that $\sigma(\sqrt{d_i}) = -\sqrt{d_i}$ for $i = 1, \ldots, l$. For any $i = l+1, \ldots, r$ we have that $d_i < 0$ in K_v, hence the image of d_i in K^*/K^{*2} can be represented as a product of an odd number of the elements $d_1 K^{*2}, \ldots, d_l K^{*2}$. Therefore, $\sigma(\sqrt{d_i}) = -\sqrt{d_i}$ for any $i = 1, \ldots, r$. Using Chebotarev's density theorem (see [6], Chapter VIII, §2.4), we get that there exist infinitely many $w \in V^K \setminus V_\infty^K$ such that L_{iw}/K_w is an unramified extension of degree 2 for all $i = 1, \ldots, r$. In particular, the group generated by the norm subgroups $N(L_{iw}/K_w)$ coincides with $U(K_w) K_w^{*2}$, where $U(K_w)$ is the group of w-adic units in K_w. Obviously, the index $[K^* : \bigcap_{j=1}^t (K^* \cap U(K_{w_j}) K_{w_j}^{*2})]$ is equal to 2^t for any collection $w_1, \ldots, w_t$ of non-Archimedean valuations of K. Therefore, it follows from what was proved that the index in K^* of the subgroup generated by $\bigcup_{i=1}^r N(L_i/K)$ is infinite, which cannot be, because $[K^* : \mathrm{Nrd}_{D/K}(D^*)] < \infty$. Proposition 1 is proved.

This proposition and its proof show that even for quaternions over number fields the question of whether $\mathrm{Nrd}_{D/K}(D^*)$ is generated by finitely many norm subgroups is nontrivial and involves the so-called multiplicative arithmetic of division algebras. The goal of this note is to get a complete answer to this question for a certain class of division algebras over number fields that includes, in particular, all division algebras of prime index.

Accordingly, let D be a division algebra of index $n = p_1^{\alpha_1} \cdots p_s^{\alpha_s}$ over a field K of algebraic numbers. Denote by n_v the index of the simple algebra $D_v = D \otimes_K K_v$ ($v \in V^K$), and let $S = \{v \in V^K \setminus V_\infty^K | n_v > 1\}$. Moreover, for $v \in V^K / V_\infty^K$ denote by q_v (respectively, π_v) the number of elements in the field of residues of K with respect to v (respectively, the uniformizing element).

THEOREM. *Assume that for all $v \in S$ the algebra D_v is a division algebra. Then $\mathrm{Nrd}_{D/K}(D^*)$ is generated by finitely many norm subgroups $N(L/K)$ corresponding to maximal subfields $L \subset D$ if and only if the following conditions hold*:
1) $n_v = 1$ for $v \in V_\infty^K$;
2) $q_v^{n_i} \equiv 1 \pmod{p_i}$ for $v \in S$ and any $i = 1, \ldots, s$ such that $p_i \nmid q_v$, where $n_i = n / p_i^{\alpha_i}$.

The finiteness of the index $[K^* : \mathrm{Nrd}_{D/K}(D^*)]$ follows from the description of $\mathrm{Nrd}_{D/K}(D^*)$ as the collection of elements in K^* that are positive with respect to all $v \in V_\infty^K$ such that $n_v > 1$ (Eichler's theorem, see [1]). Therefore, $\mathrm{Nrd}_{D/K}(D^*)$ is generated by finitely many norm subgroups if and only if D contains maximal fields $L_1, \ldots, L_d$ such that the subgroup $N(L_1/K) \cdots N(L_d/K)$ has finite index in K^*. Here it is useful to bear in mind the following assertion.

PROPOSITION 2. *Suppose that $L_1, \ldots, L_d$ are finite extensions of the field K. The subgroup $N(L_1/K) \cdots N(L_d/K)$ has finite index in K^* if and only if the subgroup $N_v(L_1/K) \times \cdots \times N_v(L_d/K)$ coincides with K_v^* for almost all $v \in V^K$.*

(Here and below if $v \in V^K$ and L/K is a finite extension, then let $N_v(L/K)$ stand for $\prod_{w|v} N(L_w/K_v)$, where the product is over all extensions of v to L. In view of the formula $N_{L/K}(x) = \prod_{w|v} N_{L_w/K_v}(x)$, $x \in L$ (see [3], Chapter XII, §3, Proposition 10), we have the imbedding $N(L/K) \subset N_v(L/K)$.)

To prove the necessity of conditions 1) and 2) in the theorem we show that if at least one of them does not hold, then for any finite collection $L_1, \ldots, L_d$ of maximal subfields of D the set $V_0 = \{v \in V^K | N_v(L_1/K) \cdots N_v(L_d/K) \neq K_v^*\}$ is infinite. This follows from the fact that in this situation there are infinitely many valuations $v \in V^K \setminus V_\infty^K$ for which all the local extensions L_{iw}/K_v ($i = 1, \ldots, d$; $w|v$) are unramified, and their degrees are divisible by a fixed prime number p. Then all the norm subgroups $N(L_{iw}/K_v)$, and hence also the group $N_v(L_1/K) \cdots N_v(L_d/K)$ are contained in the group $U(K_v) K_v^{*p} \neq K_v^*$. The existence of an infinite set of valuations with the required properties can be deduced from Chebotarev's density theorem if we apply it to a minimal normal extension L of K containing all the L_i, and to the automorphism $\sigma \in \operatorname{Gal}(L/K)$ with the following property:

$$\text{if } \sigma^a|(\tau L_i) = \operatorname{id}_{\tau L_i} \text{ for some } i = 1, \ldots, d \text{ and some } \tau \in \operatorname{Gal}(L/K), \text{ then } a \text{ is a multiple of } p, \tag{1}$$

where p is some fixed prime number. Namely, by Chebotarev's theorem, there are infinitely many valuations $v \in V^K$ such that the extension L_w/K_v is unramified, and its Frobenius automorphism φ falls in the conjugacy class $\{\tau\sigma\tau^{-1}\}$. Then all the extensions L_{iw}/K_v are also unramified, and hence are Galois extensions, and $\operatorname{Gal}(L_{iw}/K_v)$ is generated by the restriction $\varphi|L_{iw}$. Therefore, it follows from (1) that the order of $\varphi|L_{iw}$, which is equal to the degree $[L_{iw}:K_v]$, is divisible by p. Thus, the proof has been reduced to the construction of an automorphism σ satisfying (1), to which we now proceed.

It should be recalled that an extension P of K of degree n is isomorphic to a maximal subfield of D if and only if for any $v \in V^K$ and any $w|v$ the degree $[P_w:K_v]$ is divisible by n_v (see [4], §18.4). If condition 1) does not hold for $v \in V_\infty^K$, then $K_v = \mathbf{R}$ and $n_v = 2$, so that $L_{iw} = \mathbf{C}$ for any $i = 1, \ldots, d$ and any $w|v$. This implies that the restriction to L of the complex conjugation automorphism satisfies (1) for $p = 2$, and this proves the necessity of condition 1) of the theorem.

Assume now that $q_v^{n_i} \not\equiv 1 \pmod{p_i}$ for some $v \in S$ and some $i = 1, \ldots, s$ such that $p_i \nmid q_v$, where $n_i = n/p_i^{\alpha_i}$. We find the required automorphism σ already in the splitting group of the valuation v. For brevity let $\alpha = \alpha_i$, $p = p_i$, $q = q_v$, and $m = n_i$. A primitive pth root ρ_p of unity is contained in the maximal unramified extension K_v^{nr}. Let σ be the restriction to L of an automorphism $\sigma_0 \in \operatorname{Gal}(\overline{K}_v/K_v)$ such that $\sigma_0(\sqrt[p]{\pi_v}) = \rho_p \sqrt[p]{\pi_v}$, and the restriction $\sigma_0|K_v^{nr}$ generates $\operatorname{Gal}(K_v^{nr}/K_v(\rho_p))$. To prove that σ satisfies (1) it suffices to show the following:

$$\text{if } \sigma^a|(\tau L_i)_w = \operatorname{Id}_{(\tau L_i)_w} \text{ for some } i = 1, \ldots, d \text{ and some } \tau \in \operatorname{Gal}(L/K) \ (w|v), \text{ then } a \text{ is a multiple of } p. \tag{2}$$

We fix $i \in \{1, \ldots, d\}$ and $\tau \in \operatorname{Gal}(L/K)$, and denote τL_i by F. The extension F_w/K_v splits into a tower $F_w \supset P \supset E \supset K_v$, where E/K_v is a maximal unramified subextension, and P/E is a maximal weakly totally ramified subextension (see [2], Chapter II). By assumption, $[L_{iw}:K_v] = n = n_v$, therefore, the degree $[F_w:K_v]$ is also equal to n. Since $[F_w:P]$ divides some degree q, and $p \nmid q$, it follows that $[P:K_v]$ is divisible by p^α. Therefore, either $[E:K_v]$ is divisible by p, or $[E:K_v]$ divides m, but $[P:E]$ is divisible by p^α. We analyze the first case. Since $[K_v(\rho_p):K_v]|(p-1)$, it follows that $[E:E \cap K_v(\rho_p)]$ is divisible by p. But by construction, $\sigma|E$ generates $\operatorname{Gal}(E/E \cap K_v(\rho_p))$, which implies (2).

Suppose now that $[E:K_v]|m$ and $p^\alpha|[P:E]$. According to [2] (p. 38), $P = E(\theta)$, where θ is a root of the polynomial $x^a - \pi_v u$, $u \in U(E)$, and $a = [P:E]$. Let $\mu = \theta^{a/p}$. Then $\mu^p = \pi_v u$. Since $q^m \not\equiv 1 \pmod p$, it follows that, all the more so, $q^{[E:K_v]} \not\equiv 1 \pmod{p}$. But the number $q^{[E:K_v]} - 1$ coincides with the order of the multiplicative group e^* of the field of residues for E, therefore, $e^{*p} = e^*$. Using Hensel's lemma (see [2]), we get that $U(E)^p = U(E)$. In particular, $u = t^p$, $t \in U(E)$, thus $\mu = \zeta \sqrt[p]{\pi_v} t$, where ζ is some pth root of unity. If $\sigma^a|F_w = \mathrm{id}_{F_w}$, then $\sigma^a(\mu) = \mu$. On the other hand,

$$\sigma^a(\mu) = \sigma^a(\zeta \sqrt[p]{\pi_v} t) = \rho^a \mu,$$

and thus a is a multiple of p.

Let us proceed to the proof of the sufficiency of conditions 1) and 2). Since the numbers $n_i = n/p_i^{\alpha_i}$, $i = 1, \ldots, s$, are relative primes, the group K^* is generated by the groups K^{*n_i}. Therefore, it suffices to find for each $i = 1, \ldots, s$ a collection $L_1^{(i)}, \ldots, L_{d_i}^{(i)}$ of maximal subfields of D such that $N(L_1^{(i)}/K) \cdots N(L_{d_i}^{(i)}/K) \cap K^{*n_i}$ has finite index in K^{*n_i}. With this aim we construct an extension $E^{(i)}/K$ of degree n_i and extensions $L_j^{(i)}/E^{(i)}$, $j = 1, \ldots, d_i$, of degree $p_i^{\alpha_i}$ such that $L_j^{(i)} \subset D$, and the subgroup $N(L_1^{(i)}/E^{(i)}) \cdots N(L_{d_i}^{(i)}/E^{(i)})$ has finite index in $(E^{(i)})^*$. As above, in order not to over complicate the notation we fix an $i \in \{1, \ldots, s\}$ and let $p = p_i$, $\alpha = \alpha_i$, and $m = n_i$. As $E = E^{(i)}$ we take an extension of K of degree m such that all the extensions E_w/K_v ($v \in S$, $w|v$) are unramified of degree m. The possibility of such a choice follows from Krasner's lemma (see [2]; cf. [5], §2). Assume now that p is odd. By assumption, $q_v^m \equiv 1 \pmod{p}$ for all $v \in S$ such that $p \nmid q_v$, whence $U(E_w)^p \neq U(E_w)$ for all $v \in S$. Let $\zeta_v \in U(E_w)\setminus U(E_w)^p$. By the weak approximation theorem, there exist elements $a, b \in E$ such that $a \in \pi_v U(E_w)^p$ and $b \in \zeta_v U(E_w)^p$ for all $v \in S$ and $w|v$. Denote by $\{L_j\}$ a family of extensions, each generated over E by one of the roots of the polynomials of the form $x^p - ab^l = 0$ or $x^p - a^l b = 0$, where $l = 0, \ldots, p^\alpha - 1$. By using the criterion for irreducibility of polynomials of this form ([3], Chapter VIII, §8) it is not hard to show that the L_j are the desired fields.

In the case $p = 2$ the required fields L_j are obtained by α-fold application of the procedure used in Proposition 1 for a division algebra of quaternions.

The authors express deep gratitude to V. P. Platonov for the statement of the problem and for a discussion of the results obtained.

Bibliography

1. Andre Weil, *Basic number theory*, Springer-Verlag, New York, 1967.
2. Serge Lang, *Algebraic numbers*, Addison-Wesley, Reading, MA, 1964.
3. ___, *Algebra*, Addison-Wesley, Reading, MA, 1965.
4. Richard S. Pierce, *Associative algebras*, Springer-Verlag, Berlin and New York, 1982.
5. V. P. Platonov and A. S. Rapinchuk, *Multiplicative structure of division algebras over number fields and the Hasse norm principle*, Trudy Mat. Inst. Steklov. **165** (1984), 171–187; English transl. in Proc. Steklov Inst. Math. **165** (1985).

6. J. W. S. Cassels and A. Frölich (eds.), *Algebraic number theory*, Academic Press, London, and Thompson, Washington DC, 1967.

Institute of Mathematics
Academy of Sciences of the Belorussian SSR

Received 27/APR/87

Translated by H. H. McFADEN

Unitarity of the Multiplicative Group of a Group Algebra

A. BOVDI

Abstract. Let G be a group and K a commutative ring with identity. In the present paper an antiautomorphism $\sum_{g \in G} x_g g \mapsto \sum_{g \in G} x_g f(g) g^{-1}$ induced by a homomorphism of G into the multiplicative group $U(K)$ of K is considered. Those elements u in the group of units $U(KG)$ for which $u^{-1} = \varepsilon u^f$, $\varepsilon \in U(K)$, generate the f-unitary subgroup $U_f(KG) \leq U(KG)$. Interest in these subgroups has arisen in algebraic topology and in K-theory (in the case $K = \mathbf{Z}$). In the present paper the problem of characterizing groups G and fields K for which $U(KG) = U_f(KG)$ is investigated.

Let KG be the group ring of a group G over a commutative ring K with identity, and let $U(KG)$ be the multiplicative group of KG. If f is a homomorphism of G into the multiplicative group of K and $x = \sum_{g \in G} \alpha_g g$ is an element of the ring KG, then the element $x^f = \sum_{g \in G} \alpha_g f(g) g^{-1}$ is associated with x. In this case the mapping $x \mapsto x^f$ is an antiautomorphism of KG of order two and is called the *involution generated by the homomorphism* f.

An element u of the group $U(KG)$ is said to be f-unitary if the inverse element u^{-1} coincides with εu^f, where $\varepsilon \in U(K)$. It is easy to see that all the f-unitary elements of $U(KG)$ form a subgroup, which will be called the *f-unitary subgroup of* $U(KG)$ and denoted by $U_f(KG)$. In this case when $U(KG) = U_f(KG)$ the group $U(KG)$ is said to be *f-unitary*.

Interest in the group $U_f(KG)$ arose in algebraic topology and in unitary K-theory in the case when K is the ring of integers [1]. In the present article we investigate the question of the f-unitarity of the group $U(KG)$ for an arbitrary group algebra KG. In the special case when K is a finite prime

---

1991 *Mathematics Subject Classification.* Primary 20C07, 16S34.
Translation of Tartu Riikl. Ul. Toimetised No. 764 (1987), 3–11.

field the problem was treated by the author in [2], where necessary conditions for the group $U(KG)$ to be f-unitary are given. However, the author left a gap in the description of the group G.

If f is the trivial homomorphism, then the element x^f will be denoted by x^*. Let

$$V(KG) = \left\{ \sum_{g \in G} \alpha_g g \in U(KG) \Big| \sum_{g \in G} \alpha_g = 1 \right\}.$$

It is easy to see that if f is the trivial homomorphism and the element $x \in V(KG)$ is f-unitary, then $x^* = x^{-1}$.

We first present some auxiliary facts.

LEMMA 1. *Suppose that the elements of finite order in G form an Abelian subgroup $\pi(G)$, the factor group $G/\pi(G)$ is right-ordered, and the characteristic of K does not divide the orders of the elements in $\pi(G)$. If all the idempotents of the algebra $K\pi(G)$ are central in KG and $x \in U(KG)$, then there exist orthogonal idempotents $e_1, \ldots, e_n$ and elements $\alpha_i \in U(K\pi(G))$ and $g_i \in G$ such that $e_1 + \cdots + e_n = 1$ and*

$$x = \sum_{i=1}^n \alpha_i e_i g_i, \qquad x^{-1} = \sum_{i=1}^n \alpha_i^{-1} e_i g_i^{-1}.$$

PROOF. It is easy to see that the group algebra KG is isomorphic to the crossed product S of the group $G/\pi(G)$ and the group algebra $K\pi(G)$ [3]. If $x \in U(KG)$, then x and x^{-1}, as elements of S, can be represented in the form (see [3])

$$x = \sum_{i=1}^s t_{h_i} \alpha_i, \qquad x^{-1} = \sum_{j=1}^m t\overline{g}_j \beta_j \qquad (\alpha_i, \beta_j \in K\pi(G)).$$

The support subgroup of the elements α_i, β_j ($i = 1, 2, \ldots, s; j = 1, 2, \ldots, m$) is a finite Abelian normal subgroup H of G, since the idempotent $1/|H| \sum_{h \in H} h$ is central in KG. The algebra KH is semisimple, and KH contains orthogonal primitive idempotents $e_1, \ldots, e_n$ such that $e_1 + \cdots + e_n = 1$. The elements xe_k and $x^{-1}e_k$ have coefficients $\alpha_i e_k$ and $\beta_j e_k$ in the field KHe_k. Since KHe_k is invariant under transformation with the help of elements t_g ($g \in G/\pi(G)$), repetition of the arguments in the proof of Theorem 46 in [3] give us that $xe_k = t_{h_q} \alpha_q e_k$ if $xe_k \cdot x^{-1} e_k = e_k$. The lemma follows directly from this.

LEMMA 2 [2]. *Let f be a homomorphism from the group G to the field K of two elements. Then: 1) if G is a cyclic group of order 4, then $U(KG) = U_f(KG)$; 2) if G is a cyclic group of order 8, then $U(KG)$ is not f-unitary; 3) if G is the group of quaternions or the dihedral group of order 8, then the group $U(KG)$ is not f-unitary.*

LEMMA 3. *Suppose that f is the trivial homomorphism from G to a field K of characteristic p, $\langle a \rangle$ is a cyclic subgroup of finite order s in G, $g \in G$, and the group $U(KG)$ is f-unitary. In this case:*
  1) *if $s = 2$, $p = 2$, and $g^2 \neq 1$, then either $[a, g] = 1$ or $aga = g^{-1}$;*
  2) *if $s = 4$, $p = 2$, $g^2 \notin \langle a \rangle$, and $a^2 g a^2 = g^{-1}$, then this is impossible;*
  3) *if $s = 3$ and $p = 2$, then the subgroup $\langle a \rangle$ is normal in G;*
  4) *if $s = 2$ and $p = 3$, then $[a, g] = 1$.*

PROOF. Let $u = (1 + a + a^2 + \cdots + a^{s-1})g(1 - a)$. Then $u^2 = 0$ and $1 + u \in U(KG)$, and the f-unitarity of this element implies that $u + u^f = 0$. Therefore,
$$(1 + a + a^2 + \cdots + a^{s-1})g(1 - a) + (1 - a^{-1})g^{-1}(1 + a + \cdots + a^{s-1}) = 0. \quad (1)$$
We consider the cases above.

1. If $[a, g] \neq 1$, then $a \notin \langle g \rangle$, and $aga = g^{-1}$ in view of (1).

2. Since $a^2 g a^2 = g^{-1}$, it follows that $g^2 \notin \langle a \rangle$. If we multiply (1) by $1 + a$, then we get that $(1 + a + a^2 + a^3)g(1 + a^2) = 0$, which implies that $g = g^{-1}a^i$, but this is impossible.

3. Suppose that $g \notin N_G(\langle a \rangle)$. Then $u \neq 0$, and if we multiply the equality (1) by $1 + a$, then we get that $(1 + a + a^2)g(1 + a^2) = 0$, but this is impossible.

4. If $[a, b] \neq 1$, then (1) is a contradiction. The lemma is proved.

THEOREM 1. *Suppose that f is a trivial homomorphism from the group G to the field K, and the group $U(KG)$ is f-unitary. Then one of the following conditions holds:*
  1) *G is a group of exponent 2, and K is a field of characteristic 2;*
  2) *G is a cyclic group of order 4, and K is the field of two elements;*
  3) *A is a subgroup of the direct product of a torsion-free Abelian group and an elementary Abelian 3-group, G is the semidirect product of the group A and a group $\langle b \rangle$ of order 2 such that $bab = a^{-1}$ for all $a \in A$, K is a field of characteristic 2, and in the case when G has an element of order 3 the field K has two elements;*
  4) *K is the field of two elements, and in G the elements of finite order form a subgroup $\pi(G)$ such that each subgroup of $\pi(G)$ is normal in G and $\pi(G)$ is an elementary Abelian 3-group;*
  5) *K is a field of three elements, and in G the elements of finite order belong to the center and form an elementary Abelian 2-subgroup $\pi(G)$;*
  6) *G is a torsion-free group.*

If the group G and the field K satisfy one of the above conditions and, moreover, in parts 4)–6) the factor group of G by the subgroup of elements of finite order is right-ordered, then the group $U(KG)$ is f-unitary, where f is the trivial homomorphism of G into K.

PROOF. Suppose that f is the trivial homomorphism and $U(KG)$ is an f-unitary group. If a is an element of prime order q in G, and K contains a nonzero element α whose order is not equal to q, then the element $x = (a - \alpha)(1 - \alpha)^{-1}$ is invertible in KG, and

$$x^{-1} = (1 - \alpha)(1 - \alpha^q)^{-1}(\alpha^{q-1} + \alpha^{q-2}a + \cdots + \alpha a^{q-2} + a^{q-2}).$$

Then

$$xx^* = (1 + \alpha^2 - \alpha a - \alpha a^{-1})(1 - \alpha)^{-2}.$$

Suppose that the characteristic of K is different from 2. If $q > 2$, then the element $2^{-1}(a + 1)$ is not f-unitary. But if $q = 2$ and the field K has an element α whose order does not divide 4, then again the element $(a - \alpha)(1 - \alpha)^{-1}$ is not f-unitary. Consequently, if G has an element of finite order, then it follows from the f-unitarity of $U(KG)$ that either K has characteristic 2, or K is a field of three elements and all elements of G with finite order are 2-elements. We consider each case separately.

1. Suppose that K is a field of characteristic 2, and let P be a 2-Sylow subgroup of G. Then $U(KP)$ is f-unitary, the exponent of P divides 4 by Lemma 2, and the group P is locally finite by a theorem of Sanov. The group G does not have a subgroup representable as the direct product of a group $\langle a \rangle$ of order 2 and a group $\langle b \rangle$ whose order is different from 2, because the element $1 + b(a + 1)$ of order 2 is not f-unitary.

On the basis of this fact we prove that if $P \neq 1$, then P is either of exponent 2 or a cyclic group of order 4. Indeed, if P has a finite non-Abelian subgroup H, then the elements of order 2 in the center of H are contained in each cyclic subgroup of order 4 in H. If an element c of order 2 does not belong to the center of H, then in view of Lemma 3 the element c together with some element of order 4 generates the dihedral group of order 8, and this contradicts Lemma 2. Therefore, H contains a unique element of order 2, and as is known, such a group is a quaternion group of order 8. However, this is impossible, by Lemma 2, and hence P is Abelian. If P is an element of order 4, then P has a unique element of order 2 and is cyclic. We remark that if $P = \langle a | a^4 = 1 \rangle$ and the field K contains more than two elements, then it is possible to construct an element of the form $(a - \alpha)(1 - \alpha)^{-1}$ that is not f-unitary.

Let q be an odd prime number, and L a q-Sylow subgroup of G. If a is an element of order q^m in L and $q^m > 3$, then the element $x = 1 + a + a^2 + \cdots + a^{q^m - 3}$ is invertible in KL and is not f-unitary. Indeed, if $q^m \equiv 1 \pmod 4$, then $x^{-1} = a^2 + a^4 + \cdots + a^{q^m + 1}$, while if $q^m \equiv 3 \pmod 4$, then $x^{-1} = a^3 + a^5 + \cdots + a^q$. Therefore, $q^m = 3$, and, by Lemma 3, L is an elementary Abelian 3-group, and each subgroup of it is normal in G. If $H = \langle a | a^3 = 1 \rangle$, then it follows from the f-unitarity of $U(KG)$ that K is the field with two elements.

Suppose that $P \neq 1$ and the group G has an element of order 3 or of infinite order. If g is such an element, then, as mentioned above, an element a of order 2 does not commute with g, and $aga = a^{-1}$ by Lemma 3, while the group G does not contain elements of order 4. An element in the centralizer C of g has order 3 or infinite order. Therefore, $aca = c^{-1}$ for all $c \in C$, and the group C is Abelian. Suppose that $h \in G \backslash C$ and the order of h is not equal to 2. If $c \in C$ and $(ch)^2 = 1$, then we get the contradictory equality $(ch)c^{-1}(ch) = c$. Therefore, $a(hc)a = (hc)^{-1}$ for each $c \in C$, and $aha = h^{-1}$. This implies that $h \in C$, which is impossible. Consequently, the elements in $G \backslash C$ have order 2, and the index of the subgroup C is equal to 2.

We consider the second case, when K is a field of three elements and G has only 2-elements of finite order. If a is an element of order 4 in G, then $x = 1 + 2a + a^3 \in V(KG)$ and $xx^* = a^2$, but this is impossible. Consequently, the elements of finite order in G have order 2 and belong to the center of G by Lemma 3. The necessity of the conditions in the theorem is proved.

It is easy to see that if condition 1) or 2) holds, then the group $U(KG)$ is f-unitary. If 3) holds and $x \in U(KG)$, then $x = x_1 + x_2 b$ $(x_i \in KA)$ and $y = xx^* = x_1 x_1^* + x_2 x_2^* \in U(KA)$. Since A is a direct product of a torsion-free group B and a group C with exponent divisible by 3, it follows from Lemma 1 that there exist orthogonal idempotents $e_1, \ldots, e_s$ in KC such that $y = \alpha_1 e_1 g_1 + \cdots + \alpha_s e_s g_s$ $(g_i \in B,\ \alpha_i \in U(KC))$. If $C \neq 1$, then K is the field of two elements. Therefore, if $e_i = \sum_{h \in C} \alpha_h h$, then in view of the reduced Newton binomial formula

$$e_i^* = \sum_{h \in C} \alpha_h h^2 = e_i^2 = e_i.$$

It is easy to see that $y^* = \alpha_1^* e_1 g_1^{-1} + \cdots + \alpha_s^* e_s g_s^{-1}$, and the equality $y = y^*$ implies that $s = 1$ and $g_1 = 1$. Then $y^2 = y^* = y$, and $y = 1$. Therefore, the group $U(KG)$ is f-unitary along with its subgroup $U(KC)$.

If one of the conditions 4)–6) holds, then the group $U(K\pi(G))$ is f-unitary, and all the idempotents in $K\pi(G)$ are central in KG. Since the factor group $G/\pi(G)$ is right-ordered, it follows from Lemma 1 that $U(KG)$ is f-unitary. The theorem is proved.

THEOREM 2. *Suppose that K is a field of characteristic $p \geq 0$, and f is a nontrivial homomorphism of G into $U(K)$ with kernel A. If $U(KG)$ is f-unitary, then one of the following conditions holds:*

1) $G = \langle b \rangle$ *is a group of order* 2, $f(b) = -1$, *and* $p \neq 2$;

2) G *is the semidirect product of a torsion-free Abelian group A and a group $\langle b \rangle$ of order 2 such that $bab = a^{-1}$ for all $a \in A$, $f(b) = -1$, and $p \neq 2$;*

3) G *is a torsion-free group, and the field K contains more than two elements.*

If G and K satisfy one of the indicated conditions, and, moreover, in 3) *the group G is right-ordered, then the group $U(KG)$ is f-unitary.*

PROOF. Suppose that f is a nontrivial homomorphism of G into $U(K)$ with kernel A, and the group $U(KG)$ is f-unitary. Then the subgroup $U(KA)$ is unitary with respect to the involution generated by the trivial homomorphism. Therefore, the group A and the field K satisfy one of the conditions in Theorem 1, and K contains more than two elements.

Let q be a prime number, let b be an element of order q^t in the set $G\backslash A$, and let K be a field of characteristic $p \geq 0$. We prove that b is an element of order 2.

For odd q we construct the following invertible elements: 1) if $p \neq 2$, then $1+b \in U(KG)$; 2) if $p = 2$ and $q^t > 3$, then $1+b+b^2+\cdots+b^{q^{t-3}} \in U(KG)$; 3) if $p = 2$ and $q^t = 3$, and if the elements α and β of K do not belong to a prime subfield and $\alpha+\beta+\alpha\beta = 0$, then $1+\alpha(b+b^2) \in U(KG)$. It is easy to see that these elements are not f-unitary. Consequently, $q = 2$ and $p \neq 2$. Moreover, if $C = \langle b \rangle \cap \operatorname{Ker} f \neq 1$, then the group C is of order 2 and K is a field of three elements.

Suppose that $t > 1$ and K contains more than five elements. Then the group $\langle b \rangle$ has an element c of order 4, the field K has an element α whose order does not divide 4, and the invertible element $c-\alpha$ is not f-unitary; but this is impossible.

Suppose that $t > 1$, c is an element of order 4 in $\langle b \rangle$, and K is a field of three or five elements. If $p = 3$, then $x = 1+2c+c^3 \in V(KG)$, $f(c) = 2$, and $x^{-1} = 2c+c^2+c^3$, and such an element is not f-unitary. But if $p = 5$, then the element $2+3c+3c^2+3c^3$ has order 2, $f(c) \in \{2,3\}$, and again we have constructed an element that is not f-unitary. Thus, the elements of finite order in $G\backslash A$ have order 2 and belong to a single coset of G with respect to the subgroup A.

Assume that the subgroup $\langle c \rangle$ of order 2 in G does not belong to the center of G. Then $u = (1+c)g(1-c)$ is a nonzero nilpotent element for each $g \in G$ that does not commute with c. Since the element $1+u$ is f-unitary, it follows that $u+u^f = 0$, and this implies that

$$g+cg-gc-cgc+f(g)[g^{-1}+f(c)(g^{-1}c-cg^{-1})-cg^{-1}c] = 0. \quad (2)$$

Suppose that the set $G\backslash A$ has an element c of order 2. Then $p \neq 2$, $f(c) = -1$, and the centralizer $C_G(c)$ of c has order 2. Indeed, the element $x = \frac{1}{2}(1-c)a+\frac{1}{2}(1+c)$ is invertible for each $a \in C_G(c)$. It is easy to see that x is f-unitary if and only if $a \in \langle c \rangle$. Assume that $C_G(c) \neq G$. Then an element g of infinite order in $G\backslash A$ does not satisfy the equality (2), and we get on the basis of this equality that $cgc = g^{-1}$ for all $g \in A$. Consequently, $G = \langle A, c \rangle$ and $cac = a^{-1}$ for all $a \in A$, and in view of the equality $C_G(c) = \langle c \rangle$ it follows from Theorem 1 that A is a torsion-free group.

Suppose that the set $G\backslash A$ has only elements of infinite order. We prove that A is torsion-free. Indeed, otherwise the group has an element of A of order 2 in view of Theorem 1. If g is an element of infinite order in G and $[c, g] = 1$, then for characteristics $p = 2$ and $p \neq 2$ we construct the invertible elements $1 + g(1 + c)$ and $\frac{1}{2}(1 + c) + \frac{1}{2}(1 - c)g$, respectively. However, these elements are not f-unitary. Therefore, $[c, g] \neq 1$, and it follows from (2) that $g^{-1} = cgc$. Then the equality (2) is contradictory. Consequently, G is torsion-free.

We prove that the conditions obtained are sufficient for $U(KG)$ to be f-unitary if in condition 3) it is assumed in addition that G is right-ordered.

Suppose that condition 2) holds. If $u \in U(KG)$, then $u = x_1 + x_2 b$ ($x_i \in KA$), $u^f = x_1^* - x_2 b$, and $uu^f = x_1 x_1^* - x_2 x_2^* \in U(KA)$. Since $(uu^f)^* = uu^f$ and $V(KA) = A$, it follows that $\varepsilon = uu^f \in U(K)$, and the group $U(KG)$ is f-unitary.

But if G is right-ordered, then $V(KG) = G$ by Lemma 1, and hence $U(KG)$ is f-unitary. The theorem is proved.

Bibliography

1. S. P. Novikov, *The algebraic construction and properties of Hermitian analogues of K-theory over rings with involution from the point of view of the Hamiltonian formalism. Some applications to differential topology and the theory of characteristic classes.* II, Izv. Akad. Nauk SSSR Ser. Mat. **34** (1970), 475–500; English transl. in Math. USSR-Izv. **4** (1970).
2. A. A. Bovdi, *Unitarity of the multiplicative group of a group ring over a finite prime field*, Abelian Groups and Modules, Tomsk. Gos. Univ., Tomsk, 1985, pp. 11–19. (Russian)
3. ____, *Group rings*, Uzhgorod. Gos. Univ., Uzhgorod, 1974. (Russian)

Uzhgorod State University

Received 7/MAY/86

Translated by H. H. McFADEN

Profinite Completions of Linear Solvable Groups
UDC 512.54

O. I. TAVGEN'

ABSTRACT. A positive solution of Grothendieck's problem in the class of linear solvable groups over global fields is obtained.

Let G be an arbitrary residually finite group. The topology on G with the collection of all subgroups of finite index in G as a neighborhood base at the identity is called the profinite topology. Denote by $\widehat{G}$ the completion of G in the profinite topology. It is known that $\widehat{G} = \varprojlim G_i$, where G_i runs through the collection of all finite factors of G. The assignment to G of its compactification $\widehat{G}$ is a functor $\widehat{}$ from the category of residually finite groups to the category of profinite groups.

In [1] Grothendieck posed the following natural problem: if $\varphi \colon H \to G$ is a homomorphism of finitely generated residually finite groups and $\widehat{\varphi} \colon \widehat{H} \to \widehat{G}$ is an isomorphism, then is φ also an isomorphism? We note at once that if G is residually finite and $\widehat{\varphi}$ is bijective, then φ is injective, therefore, H can be assumed to be a subgroup of G.

One of the basic results in [1] is actually a positive answer to this question in the case when H is either a group of arithmetic type in the sense of the theory of group schemes, or a group of automorphisms of a finitely presented module over an arbitrary commutative ring.

In [2] a negative solution to the Grothendieck problem was obtained for the general case. Thus, the assumption that H and G are finitely generated does not ensure that φ is an isomorphism whenever $\widehat{\varphi}$ is. Also posed in [1] is the problem of finding conditions on residually finite groups (not necessarily finitely generated) under which the above question has a positive answer.

Before formulating the results we make some remarks.

REMARK 1. For residually finite groups the surjectivity of $\widehat{\varphi}$ is equivalent

---

1991 *Mathematics Subject Classification.* Primary 20E18; Secondary 19B37, 20G30.
Translation of Dokl. Akad. Nauk BSSR **31** (1987), no. 6, 485–488.

to the denseness of H in G in the profinite topology, and the injectivity of $\hat{\varphi}$ is equivalent to the profinite topology on H coinciding with the topology induced by the profinite topology from G.

REMARK 2. If G is a finitely generated solvable linear group, then, as is known, its maximal subgroups have finite index, and hence the answer to Grothendieck's question is positive by Remark 1.

REMARK 3. We need a classification of the subgroups of $\mathbf{Q}$ in explicit form (see, for example, [3]). Suppose that $L < \mathbf{Q}$ and $m\mathbf{Z} = L \cap \mathbf{Z}$. For any prime number p let $\alpha_p = \sup\{k \in \mathbf{Z} | m/p^k \in L\}$. Thus, α_p can be equal to ∞. The collection (α_p) is called the characteristic of L. Subgroups whose characteristics differ by finitely many α_p with finite values are isomorphic. Moreover, $L = \{ma/(p_1^{i_1} \cdots p_n^{i_n}) | a \in \mathbf{Z},\ 0 \le i_1 \le \alpha_{p_1}, \ldots, 0 \le i_n \le \alpha_{p_n}\}$ and $(m, p) = 1$ for p with $\alpha_p > 0$.

PROPOSITION. *Let G be a solvable group having a series $G = G_n > G_{n-1} > \cdots G_1 > G_0 = \{1\}$ such that $G_i \triangleleft G_{i+1}$, $G_{i+1}/G_i < \mathbf{Q}$, and the set of all p such that $\alpha_p = \infty$ for at least one of the characteristics of G_{i+1}/G_i does not coincide with the set of all prime numbers. In this case if $H < G$, then $\hat{H} = \hat{G}$ implies that $H = G$.*

Denote by K a global field and by V_K the set of all non-Archimedean nonequivalent valuations of K, and for $S \subset V_K$ let S_L be the restriction of S to $L \subset K$ and $\mathscr{O}_S$ the ring of S-integers of K.

THEOREM 1. *Suppose that G is an almost unipotent subgroup of $\Gamma_{\mathscr{O}_S}$, where Γ is a linear algebraic group defined over K, and $S_\mathbf{Q} \ne V_\mathbf{Q}$ in the case when $\operatorname{char} K = 0$. Then, for $H < G$, $\hat{H} = \hat{G}$ implies that $H = G$.*

Suppose now that S is finite and $\operatorname{char} K = 0$.

THEOREM 2. *If G is an almost solvable subgroup of $\Gamma_{\mathscr{O}_S}$, where Γ is a linear algebraic group defined over K, then for $H < G$ the condition $\hat{H} = \hat{G}$ implies that $H = G$.*

Platonov posed the question of the dependence of a positive solution of the Grothendieck problem for arithmetic groups G on the congruence problem. A special case of Theorem 2 gives a positive answer to this question in the class of solvable groups.

COROLLARY. *Let G be an almost solvable subgroup of $\Gamma_{\mathscr{O}_S}$, where Γ is a linear algebraic group defined over K, and suppose that the congruence problem has a positive solution for G. In this case if H is a proper subgroup of G and the congruence problem has a positive solution for H, then H cannot be dense in G in the congruence topology.*

We need some lemmas in the proof of the proposition and the theorems.

LEMMA 1. *Suppose that H is a finite group, and $G < \mathbf{Q}$. Then any extension $1 \to H \to \Gamma \xrightarrow{\varphi} G \to 1$ splits.*

We remark that $G = \varinjlim G_i$, where $G_i \cong \mathbf{Z}$. Thus, the extensions $1 \to H \to \varphi^{-1}(G_i) \xrightarrow{\varphi_i} G_i \to 1$ split. It remains only to "glue together" these extensions to form the extension φ. (By a splitting of φ_i we understand a homomorphism $\psi_i \colon G_i \to \varphi^{-1}(G_i)$ such that $\varphi_i \psi_i = \mathrm{id}$.) The set M_i of all splittings of φ_i is finite, and if $i \geq j$, then there is a natural mapping $M_i \to M_j$. It is not hard to show that the M_i with the indicated mappings form a projective system. But then $\varprojlim M_i$ is nonempty, and an element of it gives the desired splitting.

LEMMA 2. *A group satisfying the condition of the proposition is residually finite, and for each $m \in \mathbf{N}$ the number of subgroups of index m is finite.*

Both assertions of the lemma can be proved by induction on the length of a series for G. For $n = 1$ we have that $G < \mathbf{Q}$, and $G \neq \mathbf{Q}$ by assumption, consequently, if $[G \colon G_1] = m$, then $G_1 = mG$.

Suppose now that G has the series $G = G_n > G_{n-1} > \cdots > G_0 = \{1\}$. The group G_{n-1} is residually finite, $G_n/G_{n-1} < \mathbf{Q}$, and $G_n/G_{n-1} \neq \mathbf{Q}$, consequently, G_n/G_{n-1} is also residually finite. Thus, if $g \in G$ and $g \notin G_{n-1}$, then as a normal subgroup of finite index not containing g we can take the inverse image of a suitable normal subgroup of G_n/G_{n-1}. But if $g \in G_{n-1}$, then there is an $N \triangleleft G_{n-1}$, $[G_{n-1} \colon N] < \infty$, such that $g \notin N$. By the induction hypothesis, the number of subgroups of index $[G_{n-1} \colon N]$ in G_{n-1} is finite. Their intersection M will be a characteristic subgroup of G_{n-1}, hence $M \triangleleft G$ and $g \notin M$. Consider the extension $1 \to G_{n-1}/M \to G/M \to G/G_{n-1} \to 1$. By Lemma 1, it splits, hence there exists a subgroup of finite index in G/M whose inverse image in G does not contain g.

The assertion that the number of subgroups of fixed index is finite is obtained from the general fact that if G_1 and G_2 have this property, then so does the extension of G_1 by G_2.

LEMMA 3. *Suppose that $u \colon H \to G$ is a monomorphism of groups such that $\hat{u} \colon \widehat{H} \to \widehat{G}$ is an isomorphism, and let $N \triangleleft G$. Then $\hat{u}_1 \colon \widehat{H/N \cap H} \to \widehat{G/N}$ is an epimorphism. If, moreover, $\hat{u}_1$ is an isomorphism, and $\widehat{N} \to \widehat{G}$ is a monomorphism, then $\hat{u}_2 \colon \widehat{N \cap H} \to \widehat{N}$ is an epimorphism, and the topologies induced on $N \cap H$ by the profinite topologies of N and H coincide.*

The proof of Lemma 3 follows from the fact that upon completion of the commutative diagram

$$\begin{array}{ccccccccc} 1 & \longrightarrow & N \cap H & \longrightarrow & H & \longrightarrow & H/N \cap H & \longrightarrow & 1 \\ & & \downarrow u_2 & & \downarrow u & & \downarrow u_1 & & \\ 1 & \longrightarrow & N & \longrightarrow & G & \longrightarrow & G/N & \longrightarrow & 1 \end{array}$$

with exact rows we get the commutative diagram

$$\begin{array}{ccccccc} \widehat{N\cap H} & \longrightarrow & \widehat{H} & \longrightarrow & \widehat{H/N\cap H} & \longrightarrow & 1 \\ \downarrow \hat{u}_2 & & \downarrow \hat{u} & & \downarrow \hat{u}_1 & & \\ \widehat{N} & \longrightarrow & \widehat{G} & \longrightarrow & \widehat{G/N} & \longrightarrow & 1 \end{array}$$

where the rows are also exact (since $\widehat{}$ is a right-exact functor).

LEMMA 5. *If G and H satisfy the condition of the proposition, then H cannot have a series of length strictly less than for G.*

PROOF. Consider in G the series $G = G_n > G_{n-1} > \cdots > G_0 = \{1\}$. We show that $\widehat{G}_i \to \widehat{G}$ is injective. By Remark 1, it suffices to show that if $N < G_i$ and $[G_i : N] < \infty$, then there exists an $M < G$ with $[G : M] < \infty$ such that $M \cap G_i < N$. By Lemma 2, N can be assumed to be a characteristic subgroup of G_i, and hence a normal subgroup of G_{i+1}. Then by Lemma 1, the extension $1 \to G_i/N \to G_{i+1}/N \to G_{i+1}/G_i \to 1$ splits, i.e., there is a subgroup $K < G_{i+1}$, $[G_{i+1} : K] < \infty$, such that $K \cap G_i < N$. Continuing this process, we get the required subgroup M.

The injectivity of $\widehat{G}_i \to \widehat{G}$ implies the equality $\widehat{G}_i/\widehat{G}_{i-1} = \widehat{G_i/G_{i-1}}$, and the finite factors G_i/G_{i-1} are cyclic, consequently, $\widehat{G}$ has as a series $\widehat{G}_i$, where $\widehat{G}_i/\widehat{G}_{i-1}$ is a profinite cyclic group. Obviously, $\widehat{G}_i/\widehat{G}_{i-1} = \prod_{p \in M} \mathbf{Z}_p$, where M is the set of all primes such that $\alpha_p < \infty$ in the characteristic of G_i/G_{i-1}. If we fix some prime p, then the number k of those terms of the series for $\widehat{G}$ where $\mathbf{Z}_p$ is encountered is an invariant of $\widehat{G}$. Indeed, this number coincides with the number of terms of the series for a Sylow p-subgroup $\widehat{G}_p < G$, i.e., $\widehat{G}_p = \widehat{G}_p^k > \cdots > G_p^0 = \{1\}$ and $\widehat{G}_p^i/\widehat{G}_p^{i-1} = \mathbf{Z}_p$. Then, using the equality $\operatorname{cd}(\widehat{G}_p^i) = \operatorname{cd}(\widehat{G}_p^{i-1}) + \operatorname{cd}(\widehat{G}_p^i/\widehat{G}_p^{i-1})$ (see [4], §4, Proposition 22), we get that $k = \operatorname{cd}(\widehat{G}_p)$.

We remark now that by assumption there exists a p such that $\mathbf{Z}_p$ is encountered in each $\widehat{G}_i/\widehat{G}_{i-1}$. But since $\widehat{H} = \widehat{G}$, H cannot have a series of smaller length.

The proof of the proposition is obtained by induction on the length n.

a) For $n = 1$ we have that $G < \mathbf{Q}$. Then, by using the equality $\widehat{H} = \widehat{G}$ and Remark 1, we can show that the characteristics of H and G coincide, and $\mathbf{Z} \cap H = \mathbf{Z} \cap G$. From this, by Remark 3, $H = G$.

b) The group G has the series $G = G_n > \cdots > G_0 = \{1\}$. Consider the following commutative diagram:

$$\begin{array}{ccccccccc} 1 & \longrightarrow & H \cap G_{n-1} & \longrightarrow & H & \longrightarrow & H/H \cap G_{n-1} & \longrightarrow & 1 \\ & & \downarrow u_2 & & \downarrow u & & \downarrow u_1 & & \\ 1 & \longrightarrow & G_{n-1} & \longrightarrow & G & \stackrel{\psi}{\longrightarrow} & G/G_{n-1} & \longrightarrow & 1 \end{array}$$

By Lemma 3, upon completion we get that $\hat{u}_1$ is an epimorphism. But $\hat{u}_1$ is bijective. Indeed, if $\hat{u}_1$ is not injective, then $H/H \cap G_{n-1}$ contains a collection of subgroups H_i with $[(H/H \cap G_{n-1}): H_i] = p^{k_i}$ and $H_{i+1} < H_i$, where $k_i \to \infty$ as $i \to \infty$. Consequently, $[H: (\psi|_H)^{-1}(H_i)] = p^{k_i}$ and $H \cap G_{n-1} < (\psi|_H)^{-1}(H_i)$. But $\hat{u}$ is an isomorphism, hence there exist $\Gamma_i < G$ such that $\Gamma_i \cap H = (\psi|_H)^{-1}(H_i)$ (Remark 1). The series $\psi(\Gamma_i)$ stabilizes at some place by the choice of H_i, therefore, $[G_{n-1}: (G_{n-1} \cap \Gamma_i)] = p^{l_i}$, where $l_i \to \infty$ as $i \to \infty$, and, moreover, $H \cap G_{n-1} < \Gamma_i \cap G_{n-1}$. It is not complicated to show that in this case $H \cap G_{n-1}$ has a series of length strictly less than for G_{n-1}, and hence H has a series of length strictly less than for G. But this contradicts Lemma 5. Thus, $\hat{u}_1$ is an isomorphism, and $\psi(H) = G/G_{n-1}$. The proof that $\widehat{H \cap G_{n-1}} \to \widehat{H}$ is injective is analogous to the proof that $\widehat{G}_{n-1} \to \widehat{G}$ is injective in Lemma 5. Then we get the bijectivity of $\hat{u}_2$ by using Lemma 3. By the induction hypothesis, this implies that u_2 is an isomorphism, and the commutativity of the diagram immediately implies that $H = G$.

PROOF OF THEOREM 1. The group G will be assumed to be a subgroup of $GL_n(\mathcal{O}_S)$. In view of Lemma 3, G can be assumed to be unipotent. In this case it can, as is known, be reduced over K to unitriangular form.

If $\operatorname{char} K = p > 0$, then G has a series whose factors are periodic Abelian groups of exponent p. Hence, the condition implies that $H = G$.

Suppose now that $\operatorname{char} K = 0$. We use the standard construction of restriction of the ground field. Considering the restriction $S_\mathbf{Q} \neq V_\mathbf{Q}$ and the fact that, in the denominators of the elements of a matrix reducing G to unipotent form, prime numbers occur to a finite power, we get that G has the same form as in the condition of the proposition, and hence $H = G$.

PROOF OF THEOREM 2. Using the Lie-Kolchin theorem, we reduce G over $\overline{K}$ to almost triangular form. Adjoining to K the elements of the reducing matrix, we get a larger field K', and G is almost triangular in the group $GL_n(\mathcal{O}_{S'})$, where S' is a finite subset of $V_{K'}$. Let $T \triangleleft G$ be such that $[G: T] < \infty$ and $T < T_n(\mathcal{O}_{S'})$. Then, using Lemma 3, we see that it suffices to prove the theorem in the situation $H \cap T < T$. If $\operatorname{char} K = 0$, then, applying to T the construction of restriction of the ground field, we get by the finiteness of S' that G satisfies the condition of the proposition, and then $H = G$.

In the proposition and Theorem 1 it is impossible to do without the condition on the characteristics of the factors in the first case, and without the condition $S_\mathbf{Q} \neq V_\mathbf{Q}$ in the second case if $\operatorname{char} K = 0$.

Theorem 2 cannot be strengthened by allowing an infinite S.

In conclusion the author expresses deep gratitude to V. P. Platonov for his undivided attention to this work and for his advice. The author also thanks A. S. Rapinchuk for useful discussions.

Bibliography

1. Alexander Grothendieck, *Representations lineaires et compactification profinie des groupes discrets*, Manuscripta Math. **2** (1970), 375–396.
2. V. P. Platonov and O. I. Tavgen', *On the Grothendieck problem of profinite completions of groups*, Dokl. Akad. Nauk SSSR **288** (1986), no. 5, 1054–1058; English transl. in Soviet Math. Dokl. **33** (1986).
3. A. G. Kurosh, *Theory of groups*, 3rd ed., "Nauka", Moscow, 1967; English transl. of 2nd ed., I, II, Chelsea, New York, 1955, 1956, 1960.
4. Jean-Pierre Serre, *Cohomologie galoisienne*, 3rd ed., Lecture Notes in Math., vol. 5, Springer-Verlag, Berlin and New York, 1965.

Institute of Mathematics
Academy of Sciences of the Belorussian SSR

Received 2/JUNE/86

Translated by H. H. McFADEN

On the Grothendieck and Platonov Conjectures

UDC 512.54

O. I. TAVGEN'

Abstract. The connection between Grothendieck's and Platonov's hypotheses is established.

Let G be a group. Denote by $\widehat{G}$ the profinite completion of G, i.e., the completion of G in the topology with all subgroups of finite index in G as a neighborhood base at the identity. In [1] Grothendieck formulated the following problem.

Let $\varphi: G_1 \to G_2$ be a homomorphism of finitely generated residually finite groups such that $\hat{\varphi}: \widehat{G}_1 \to \widehat{G}_2$ is an isomorphism. Is φ also an isomorphism? Find conditions on G_1 and (or) G_2 such that φ is an isomorphism if $\hat{\varphi}$ is an isomorphism.

As shown in [2], in the general case the Grothendieck problem has a negative solution. In the counterexample constructed G_2 has the form $F_n \times F_n$, $n \geq 4$. Since $F_n \times F_n$ can be realized in $SL_2(\mathbf{Z}) \times SL_2(\mathbf{Z})$ as an arithmetic subgroup, and since the congruence problem has a negative solution for it, this leads to a natural problem formulated by Platonov in [2]: the problem of the dependence of the Grothendieck problem for arithmetic groups on a positive solution of the congruence problem. We have in view the following problem: if G_2 is an arithmetic group with a positive solution of the congruence problem, then does Grothendieck's question have a positive answer for it?

We remark that under the conditions of the Grothendieck problem G_1 can be assumed to be a subgroup of G_2. For brevity it will be assumed that the Grothendieck problem has a positive solution for G_2 if it is true for any G_1, and similarly for G_1.

1991 *Mathematics Subject Classification.* Primary 20G15, 20E18.
Translation of Dokl. Akad. Nauk BSSR **32** (1988), no. 6, 489–492.

On the other hand, Platonov proposed a deep conjecture about the connection between representations of a group and the property of it being arithmetic. We recall some concepts. Let k be a field of characteristic 0. We consider the set $\text{Hom}(G, GL_n(k))$ of representations of degree n over k for a finitely generated group G. Corresponding to each such representation is a point in $\prod_{i=1}^{m} GL_n(k)$, where m is the number of generators of G. Consequently, $\text{Hom}(G, GL_n(k))$ is the set of k-points of the affine variety determined in the natural way by the relations of G. We remark that an isomorphic manifold is obtained in passing to another system of generators.

On the given manifold of representations we can consider the subset of regular functions that consists of the functions of the form $\tau_g \colon \rho \mapsto \text{tr}\,\rho(g)$, $\rho \in \text{Hom}(G, GL_n(k))$. The algebra over k generated by these functions is finitely generated (see [3]). We choose generators $\text{Hom}(G, GL_n(k)) \to k^s \colon \rho \mapsto (\tau_{g_1}(\rho), \ldots, \tau_{g_s}(\rho))$. It can be proved to have closed range in k^s. This range is denoted by $X_n^k(G)$ and is called the manifold of characters.

Then the Platonov conjecture asserts that if G is a finitely generated group and $\dim X_n^k(G) = 0$ for any n and k, then G is isomorphic to an arithmetic group. It is easy to see that the condition $\dim X_n^k(G) = 0$ means that the number of equivalence classes of irreducible degree-n representations of G over k is finite; the converse is also true. Therefore, the conjecture can be reformulated in direct terms: if for any n and k the number of equivalence classes of irreducible representations is finite, then G is an arithmetic group. By an arithmetic group we mean a group commensurable with the group of S-units of some arithmetic group over a global field K.

In the present article we investigate the connection between the Grothendieck problem and the Platonov conjectures for finitely generated linear groups. The main result is presented in Theorem 2.

For convenience of references we number the conjectures (1), (2), and (3) in the order of their formulation in this paper.

We first present some assertions needed below. Denote by A a commutative right with identity, and by $\text{Rep}_A(G)$ the category of representations of G over finitely presentable modules M over A.

THEOREM 1. *Suppose that $\varphi \colon G_1 \to G_2$ is a homomorphism of finitely generated groups such that $\hat{\varphi} \colon \widehat{G}_1 \to \widehat{G}_2$ is an isomorphism. Then the functor $\text{Rep}_A(G_2) \to \text{Rep}_A(G_1)$ is an equivalence of categories.*

COROLLARY 1. *Under the conditions of Theorem 1 the natural mapping $\text{Hom}(G_2, \text{Aut}_A(M)) \to \text{Hom}(G_1, \text{Aut}_A(M))$ is bijective.*

All these are results of Grothendieck proved in [1].

LEMMA 1. *Suppose that $G_1 < G_2$ and $\varphi \colon G_1 \to G_2$ is the canonical imbedding. Then*:

a) $\hat{\varphi}$ *is surjective if and only if* $[G_2 : \Gamma_2] = [G_1 : (G_1 \cap \Gamma_2)]$ *for any* $\Gamma_2 < G_2$ *with* $[G_2 : \Gamma_2] < \infty$;

b) $\hat{\varphi}$ *is injective if and only if for any* $\Gamma_1 < G_1$, $[G_1 : \Gamma_1] < \infty$, *there is a* $\Gamma_2 < G_2$, $[G_2 : \Gamma_2] < \infty$, *such that* $\Gamma_2 \cap G_1 = \Gamma_1$.

LEMMA 2. *Suppose that conjecture* (1) *is valid for* G_2 (G_1). *In this case if* $[\Gamma_2 : G_2] < \infty$ ($[\Gamma_1 : G_1] < \infty$), *then conjecture* (1) *is also valid for* Γ_2 (Γ_1).

We prove Lemma 2 for G_1 (the proof is analogous for G_2). Let $\varphi \colon \Gamma_1 \to H$, and let $\hat{\varphi}$ be an isomorphism. According to Lemma 1b), there exists an $H_1 < H$ with $[H : H_1] < \infty$ such that $H_1 \cap \Gamma_1 = G_1$. Then for $G_1 < H_1$ we have that $\widehat{G}_1 = \widehat{H}_1$. The proof of surjectivity is obtained by using a) in Lemma 1, and the proof of injectivity is obtained by using b). Then $G_1 = H_1$ by the proposition, and hence $\Gamma_1 = H$.

LEMMA 3. *Suppose that* $\varphi \colon G_1 \to G_2$ *is a homomorphism of finitely generated groups, and* $\hat{\varphi}$ *is an isomorphism. Moreover, suppose that* G *is a* k-*defined algebraic group*, A *is a* k-*algebra, and* B *is a subring of* A. *Then the mapping* $\varphi_{G(B)} \colon \mathrm{Hom}(G_2, G(B)) \to \mathrm{Hom}(G_1, G(B))$ *is bijective.*

The lemma is actually a corollary to Theorem 1. We assume that $G(B) < GL_n(B)$. Then the injectivity of $\varphi_{G(B)}$ is obtained immediately from Corollary 1. To prove that $\varphi_{G(B)}$ is surjective it is necessary to show that for any homomorphism $f \colon G_1 \to G(B)$ there exists a homomorphism $g \colon G_2 \to G(B)$ such that $f = g \circ \varphi$. Since G_1 and G_2 are finitely generated, B can be assumed to be finitely generated over $\mathbf{Z}$. Then each maximal ideal in B has finite index. Moreover, the nilradical of B coincides with the intersection of the maximal ideals and is nilpotent (see [4], Chapter 5, §3). Consequently, $\bigcap_{m \in \mathrm{Max}} \bigcap_{n \in \mathbf{N}} m^n = 0$, where Max is the collection of maximal ideals in B.

By Corollary 1, there exists a homomorphism $g_1 \colon G_2 \to GL_n(B)$ such that $f = g_1 \circ \varphi$. Assume that $\mathrm{Im}\, g_1 \not\subset G(B)$. Then there exists a matrix $M \in \mathrm{Im}\, g_1$ whose elements do not satisfy one of the equations defining G, for example, $h(M) \neq 0$. By adjoining to B finitely many elements if necessary, we can assume that the coefficients in h lie in B. Then, as mentioned above, there exists a maximal ideal $m \subset B$ such that $h(M) \notin m^n$. We reduce modulo m^n. Then $\widetilde{h}(\widetilde{M}) \neq 0$, but since $\hat{\varphi}$ is an isomorphism, it follows that $\mathrm{Im}\, g_1 = \mathrm{Im}\, f < G(B/m^n B)$. Contradiction. Consequently, $\mathrm{Im}\, g_1 < G(B)$, and g_1 can be taken as g.

THEOREM 2. *Suppose that* $\varphi \colon G_1 \to G_2$ *is a homomorphism of linear finitely generated groups, and that for one of the* G_i *the equality* $\dim X_n^k(G_i) = 0$ *holds for arbitrary* n *and* k. *Then the validity of conjecture* (3) *implies that of conjecture* (1).

PROOF. Suppose that $\dim X_n^k(G_2) = 0$ for G_2. Choosing generators in G_2 such that a subset of them generates G_1, we use Corollary 1. We have the following morphism of affine manifolds: $\varphi_{GL}^{-1}\colon \operatorname{Hom}(G_1, GL_n(k)) \to \operatorname{Hom}(G_2, GL_n(k)) \to k^s\colon \rho \mapsto (\tau_{g_1}(\varphi_{GL}^{-1}(\rho)), \ldots, \tau_{g_s}(\varphi_{GL}^{-1}(\rho)))$.

But φ_{GL}^{-1} is bijective, and the range of the composition mapping is finite, consequently, so is the range of $\operatorname{Hom}(G_1, GL_n(k)) \to k^l\colon \rho \mapsto (\tau_{g_1}(\rho), \ldots, \tau_{g_l}(\rho))$, where $\tau_{g_1}, \ldots, \tau_{g_l}$ are generators of the Fricke character algebra of G_1. But this range is $X_n^k(G_1)$. Hence, G_1 also has only finitely many irreducible representations for fixed n and k. By conjecture (3), G_1 is then an arithmetic group. Thus, there exists a K-defined algebraic group G such that $G_1 \cap G(\mathcal{O}_S)$ has finite index in G_1 and $G(\mathcal{O}_S)$. According to Lemma 3, the mapping $\operatorname{Hom}(\Gamma_2, G(\mathcal{O}_S)) \to \operatorname{Hom}(\Gamma_1, G(\mathcal{O}_S))$ is bijective for any homomorphism $u\colon \Gamma_1 \to \Gamma_2$ of finitely generated residually finite groups under the condition that $\hat{u}\colon \widehat{\Gamma}_1 \simeq \widehat{\Gamma}_2$. Then $\operatorname{Hom}(\Gamma_2, G_1 \cap G(\mathcal{O}_S)) \to \operatorname{Hom}(\Gamma_1, G_1 \cap G(\mathcal{O}_S))$ is also bijective. Indeed, it suffices to prove that for any mapping $f\colon \Gamma_1 \to G_1 \cap G(\mathcal{O}_S)$ there is a homomorphism $g\colon \Gamma_2 \to G_1 \cap G(\mathcal{O}_S)$ with the condition $f = g \circ u$. But there exists a $g_1\colon \Gamma_2 \to G(\mathcal{O}_S)$ such that $f = g_1 \circ u$. We consider $g_1(\Gamma_2) \cap (G_1 \cap G(\mathcal{O}_S))$. This is a subgroup of finite index in $g_1(\Gamma_2)$, and if it is not equal to $f(\Gamma_2)$, then $f^{-1}(g_1(\Gamma_2)) \cap (G_1 \cap G(\mathcal{O}_S))$ is a proper subgroup of finite index in Γ_2 that contains Γ_1. But this contradicts Lemma 1a). Hence,

$$g_1(\Gamma_2) \cap (G_1 \cap G(\mathcal{O}_S)) = f(G_1),$$

i.e., $g_1(\Gamma_2) < G_1 \cap G(\mathcal{O}_S)$.

We now take Γ_1 to be the group $G_1 \cap G(\mathcal{O}_S)$ and consider the inverse image of the identity homomorphism under the mapping $u_{G_1 \cap G(Q_s)}$. This is the homomorphism $\Gamma_2 \to G_1 \cap G(\mathcal{O}_S)$ with the condition that the composition $G_1 \cap G(\mathcal{O}_S) \to \Gamma_2 \to G_1 \cap G(\mathcal{O}_S)$ is the identity. It follows from this that $\Gamma_2 \to G_1 \cap G(\mathcal{O}_S)$ is surjective, and, passing to the completions, we get an epimorphism $\widehat{\Gamma}_2 \to \widehat{G_1 \cap G(\mathcal{O}_S)} \cong \widehat{\Gamma}_2$. But this homomorphism is bijective, because finitely generated profinite groups are Hopfian. Consequently, using the residual finiteness of Γ_2 and $G_1 \cap G(\mathcal{O}_S)$, we get that $G_1 \cap G(\mathcal{O}_S) = \Gamma_2$. From this, $G_1 = G_2$ by Lemma 2.

COROLLARY 2. *Suppose that G_2 is a K-semisimple algebraic group, and the congruence kernel is finite for $G_2(\mathcal{O}_S)$. In this case if the conjecture (3) is valid, then so is the conjecture (2).*

It suffices to show that $G_2(\mathcal{O}_S)$ has finitely many irreducible representations for fixed n and k, and then to use Theorem 2.

As shown in [5], in this case each representation is "almost algebraic", and hence completely reducible. But then there are finitely many such representations by Proposition 2.9 in [6].

We turn now to the dependence which is converse to the corollary of Theorem 2. Here we need the following result of Rapinchuk.

LEMMA 4. *Suppose that $\rho: \Gamma \to GL_n(k_0)$ is a completely reducible finitely generated group and* $\dim X_n^k(\Gamma) = 0$ *for any n and k. Then Γ can be assumed to be a subgroup $GL_m(\mathbf{Z}_S)$, where S is finite.*

Since the finiteness of the number of classes of irreducible representations means the finiteness of the trace mappings, all the traces of the matrices in Γ belong to $k_0 \cap \overline{\mathbf{Q}}$, i.e., are algebraic. Indeed, $\Gamma < GL_n(\mathscr{O})$, where $\mathscr{O}$ is a finitely generated ring. If $A \in \Gamma$ and $\operatorname{tr}(A) = t \notin k_0 \cap \overline{\mathbf{Q}}$, then according to [4] (§3 in Chapter 5), there exists an $f \in \mathbf{Z}[t]$ such that if $\alpha \in \overline{\mathbf{Q}}$ is not a root of f, then $\varphi_\alpha(t) = \alpha$ for some homomorphism $\varphi_\alpha: \mathscr{O} \to \overline{\mathbf{Q}}$. Thus, it is possible to get a representation ρ_α such that $\operatorname{tr}(\rho_\alpha(A)) = \alpha$. Contradiction.

The Fricke characters $\tau_g: \rho \mapsto \operatorname{tr} \rho(g)$ are regular functions. Consequently, we have the equations $\operatorname{tr} \rho(g_i) = 0$ and the equations arising from the relations in Γ. These equations have solutions in $\overline{k}_0$ and coefficients in $\mathbf{Q}$, hence they have solutions also in $\overline{\mathbf{Q}}$. Consequently, $\Gamma < GL_n(\mathscr{O}) < GL_n(\overline{\mathbf{Q}})$. Using next the construction of restriction of the ground field, we get the required result.

Now let Γ be a completely reducible finitely generated subgroup([1]) of $GL_n(k_0)$, and let $\dim X_n^k(\Gamma) = 0$. Then by Lemma 4, we regard Γ as a subgroup of $GL_m(\mathbf{Z}_S)$. We consider the closure $\widetilde{\Gamma}$ of Γ in the Zariski topology. Since we are concerned with conjecture (2), we assume that Γ and $\widetilde{\Gamma}(\mathbf{Z}_S)$ have trivial congruence kernels. Under these assumptions we have the following result.

PROPOSITION 2. *If $\widetilde{\Gamma}$ is a $\mathbf{Q}$-defined group, then the validity of conjecture (2) implies that of conjecture (3) for Γ.*

It is actually possible to obtain the following

PROPOSITION 2'. *Suppose that G is a $\mathbf{Q}$-defined semisimple group, and the congruence conjecture and conjecture (2) are valid for $G(\mathbf{Z}_S)$. In this case if $\Gamma < G(\mathbf{Z}_S)$ is dense in the Zariski topology and the congruence conjecture is valid for Γ, then the index of Γ in $G(\mathbf{Z}_S)$ is finite.*

The author takes it as his pleasant duty to express gratitude to V. P. Platonov for useful discussions and for advice in the writing of this paper.

BIBLIOGRAPHY

1. Alexander Grothendieck, *Representations lineaires et compactification profinie des groupes discrets*, Manuscripta Math. **2** (1970), 375-396.
2. V. P. Platonov and O. I. Tavgen', *On the Grothendieck problem of profinite completions of groups*, Dokl. Akad. Nauk SSSR **288** (1986), no. 5, 1054-1058; English transl. in Soviet Math. Dokl. **33** (1986).
3. C. Procesi, *The invariant theory of $n \times n$ matrices*, Adv. Math. **19** (1976), no. 3, 306-381.

([1])The author has been able to show that the condition of complete reducibility of the group Γ can be dropped.

4. N. Bourbaki, *Algèbre commutative*, Chaps. 1-7, Actualités Sci. Indust., nos. 1290, 1293, 1308, 1314, Hermann, Paris, 1961, 1964, 1965.
5. M. S. Raghunathan, *On the congruence subgroup problem*, Inst. Hautes Études Sci. Publ. Math. **46** (1976), 107-161.
6. A. Lubotzky and A. Magid, *Varieties of representations of finitely generated groups*, Mem. Amer. Math. Soc. **58** (1985), no. 336, 1-117.

Institute of Mathematics
Academy of Sciences of the Belorussian SSR

Received 26/AUG/87

Translated by H. H. McFADEN

The Congruence Problem for Solvable Algebraic Groups over Global Fields of Positive Characteristic

UDC 512.743

A. A. SHAROMET

> Abstract. Let k be a global field, $\operatorname{char} k > 0$, and S be any finite set of valuations of k, $|S| > 1$. It is proved that if G is a torus or a k-solvable linear algebraic group, Cartan subgroups of which is a torus, then every S-arithmetic subgroup of G is an S-congruence.

Let G be a connected linear group defined over a global field k, and let S be a finite set of nonequivalent valuations of k that contains all the infinite valuations. Denote by O_S the ring consisting of the elements of k that are integers with respect to all valuations not in S, and by A_S its multiplicative group. Then two topologies can be introduced on the group $G(O_S)$ by choosing the neighborhood base for the identity to be the subgroups of finite index in $G(O_S)$ or to be the congruence subgroups. The congruence problem consists in computing the kernel $c(G(S))$ of the natural homomorphism $\widehat{G(O_S)} \to \overline{G(O_S)}$, extending the identity mapping to the completions of $G(O_S)$ in these topologies, or, more narrowly, in answering the question of when $c(G(S))$ is trivial. The triviality of the congruence kernel $c(G(O_S))$ is equivalent to each subgroup of finite index in $G(O_S)$ containing a congruence subgroup. All the foregoing applies to any subgroup of $GL(n, O_S)$. More detailed information about the congruence problem is contained in [1]. We remark only that in this survey Raghunathan asserts without proof that $c(T(S))$ is nontrivial if T is a torus defined over a field of positive characteristic. Our Theorem 1 refutes this assertion; its proof is on the whole analogous to an argument of Chevalley in [2], where he proved the triviality of the congruence kernel for a torus defined over a field of algebraic

---

1991 *Mathematics Subject Classification*. Primary 20G30; Secondary 19B37.
Translation of Dokl. Akad. Nauk BSSR **31** (1987), no. 3, 201–204.

©1992 American Mathematical Society
0065-9290/92 $1.00 + $.25 per page

numbers. In [3] Platonov proved the triviality of the congruence kernel for a solvable linear group over $\mathbf{Z}$, and in [4] he solved jointly with the author the congruence problem for such groups over an arbitrary ring of S-integers in $\mathbf{Q}$.

The purpose of the present article is to study the problem for solvable algebraic groups defined over fields of algebraic functions of a single variable with a finite field of constants, i.e., over global fields of positive characteristic p. In what follows it will be assumed that $|S| > 1$.

Consider first the case of a unipotent group. Here the key role is played by the following simple remark.

LEMMA 1. *An Abelian group of countable rank whose exponent is a prime number has a continuum of subgroups of finite index.*

PROOF. A group G with countable rank and prime exponent p is a direct product of a countable set of cyclic groups C_p of order p. It is not hard to see that there is a continuum of homomorphisms $G \to C_p$, but there are only finitely many homomorphisms with a single kernel. Consequently, there is a continuum of subgroups of index p.

We consider now an arbitrary infinite unipotent linear group over the ring O_S of characteristic p.

PROPOSITION 1. *The congruence kernel is nontrivial for an infinite unipotent group U of matrices over the ring O_S of characteristic p.*

PROOF. We refine the lower central series of U so that all its factors become groups of exponent p: $U = U_1 \supset U_2 \supset \cdots \supset U_k \supset U_{k+1} = \{e\}$.

Let i_0 be the minimal index such that the factor group U_{i_0}/U_{i_0+1} is infinite. Then there is a continuum of subgroups of finite index in this factor group, and hence in the group U. The proof of the proposition follows from the fact that there is a countable set of congruence subgroups.

REMARK. A local ring of characteristic p can be taken in Proposition 1 in place of the ring O_S.

We now consider the congruence problem for tori. The problem can be reduced in the standard way to the case of a one-dimensional split torus, and in this case the triviality of the congruence kernel is a consequence of the following theorem.

THEOREM 1. *For any positive integer m there is a congruence subgroup of A_S that is contained in A_S^m.*

PROOF. The proof of Theorem 1 is analogous to the main result of Chevalley in [4] and differs essentially from his arguments only in the case when m is divisible by $p = \operatorname{char} k$.

It is not hard to see that $(k^*)^m \cap A_S = A_S^m$, and hence it suffices to find in A_S a congruence subgroup whose elements have mth roots in k.

We show that in the proof of the theorem we can confine ourselves to the case when m is a power of a prime number. For this it suffices to note that if there exist $y_1, y_2 \in k$ such that $x = y_1^\alpha = y_2^\beta$ and $(\alpha, \beta) = 1$, then there exists a $y \in k$ such that $x = y^{\alpha\beta}$; indeed, $y = y_1^v y_2^u$, where $\alpha u + \beta v = 1$.

The following assertion enables us to assume in our arguments that the field k contains a primitive mth root of unity, $(m, p) = 1$.

LEMMA 2. *Suppose that ε is a primitive root of unity of degree $m = q^\alpha$, where q is a prime number, $q \neq p$, and suppose that an element $x \in k$ has an mth root in the field $k(\varepsilon)$.*

Then x also has an mth root in the field k.

The lemma can be proved by induction on the number α. It must be proved separately that -1 can be assumed to be a square in k for $m = 2^\alpha$. The proof repeats the arguments of Chevalley in [2] almost word-for-word.

It will be assumed that k contains a primitive root of unity of degree $m = q^\alpha$. Let L be the field obtained by adjoining to k the mth roots of the elements in the group A_S. This is an Abelian extension of k of finite degree, since k contains a primitive mth root of unity, and A_S is finitely generated. For each $\sigma \in G(L/k)$ there exist infinitely many unramified valuations w of the field L such that σ is the Frobenius automorphism corresponding to w (see [5], p. 384). By virtue of this we can choose for each $\sigma_i \in G(L/k)$ a valuation w_i lying over a $v_i \notin S$ and having the property indicated above. Now take the ideal $I = \{a \in O_S | v_i(a) < 1\}$ of the ring O_S, and let $x \in A_S$, $x \equiv 1 \pmod{I}$. Then $k(\sqrt[m]{x}) = k$, since $k_{v_i}(\sqrt[m]{x}) = k_{v_i}$, because $|x|_{v_i} < 1$, and thus $\sigma_i(\sqrt[m]{x}) = \sqrt[m]{x}$ by the choice of v_i. Therefore, A_S^m contains a congruence subgroup modulo I.

We now consider the case $m = p^\alpha$, where $p = \operatorname{char} k$. Let $v \notin S$ be a valuation of k, and let $\pi \in k$ be a uniformizing element for it. We imbed the group A_S in the group A_v of units of the ring of integers of the field k_v. Denote by B its closure in A_v. Then B is isomorphic to the direct product of a finite group with order relatively prime to p and finitely many copies of the group $\mathbf{Z}_p$. This implies that B^{p^α} is open in the group B with topology induced by the congruence topology of the group A_v, and hence $B^{p^\alpha} \supset B \cap A_v(\pi^t)$ $(A_v(\pi^t) = \{x \in A_v | x \equiv 1 \pmod{\pi^t}\})$. This implies that any element $x \in A_S \cap A_v(\pi^t)$ has a p^αth-root in the field k_v, but then it has such a root in the field k, because k_v does not contain elements that are not separable over k, and this concludes the proof of the theorem.

Let us now proceed to the study of the congruence problem for k-solvable algebraic groups. We need the following

LEMMA 3. *Let $G = M \ltimes N$ be a semidirect product, where everything is defined over k, and the group G is connected.*

Then $c(G(S)) = \{e\}$ if and only if $c(M(S)) = \{e\}$ and every arithmetic subgroup of $N(O_S)$ invariant under $M(O_S)$ contains a congruence subgroup.

The proof coincides in essence with that of Lemma 2 in [3].

PROPOSITION 2. *Suppose that* $G = G_m \ltimes G_a$, *where the action of* G_m *on* G_a *is defined by the character* $x \to x^m$, $m \neq 0$.
Then $c(G(S)) = \{e\}$.

PROOF. By Theorem 1 and Lemma 3, it suffices to prove that each subgroup of finite index in $G_a(O_S) = O_S$ and invariant under the action of $G_m(O_S) = A_S$ contains a congruence subgroup. But such a subgroup is invariant under multiplication by elements in A_S^m, and hence is a module over the ring $F_q[A_S^m]$, which contains the Dedekind ring A. Since $[O_S : U] < \infty$, for these two A-modules there is a $y \in A$ such that $yO_S \subset U$, i.e., U contains a congruence subgroup. To conclude the proof it remains to observe that $c(G_m(S)) = \{e\}$ in view of Theorem 1.

This result is not hard to extend to a semidirect product of a k-split torus and the additive group of a vector space if the torus acts nontrivially.

PROPOSITION 3. *Let* $G = D \ltimes V$ *be a* k-*solvable group, where* D *is a* k-*torus, and* V *is the additive group of a vector space, on which* D *acts without fixed points.*
Then $c(G(S)) = \{e\}$.

We now prove a more general result for k-solvable groups.

THEOREM 2. *Suppose that* $G = T \ltimes U$ *is a semidirect product* (*over* k) *of a* k-*split torus and a unipotent* k-*group* U, *and the torus acts on* U *without fixed points.*
Then $c(G(S))$ *is trivial.*

PROOF. By Theorem 1 and Lemma 3, it suffices to prove that every $T(O_S)$-invariant subgroup of infinite index in $U(O_S)$ contains a congruence subgroup. Note that the conditions of the theorem imply that U is k-solvable.

We refine the upper central series of the group U so that all its factors are additive groups of vector spaces: $U = U_1 \supset U_2 \supset \cdots \supset U_k \supset U_{k+1} = \{e\}$. Then the variety of U is T-equivariantly isomorphic over k to the direct product V of the factors of this series. This is ensured by the fact that the groups U_i are k-solvable, and hence the factors with respect to them have regular sections. It is not hard to see that the indicated isomorphism τ of the manifolds and its inverse are continuous in the congruence topology.

Hence, it suffices for us to show that the image of any S-arithmetic subgroup H of $U(O_S)$ invariant under $T(O_S)$ is an S-arithmetic subgroup of V. Indeed, the image of H is invariant under $T(O_S)$, and by Lemma 3 and Proposition 3, it contains a congruence subgroup if $\tau(H)$ is S-arithmetic; but then its complete inverse image, i.e., H, also contains a congruence subgroup.

Accordingly, let H be a subgroup of finite index in $U(O_S)$ that is invariant under $T(O_S)$. We show that $\tau(H) < V(O_S)$ is also arithmetic. The mapping τ is not a group homomorphism, but its restriction to U_k is a k-isomorphic imbedding of the group U_k in V, and moreover, if

$\tau(u) = (\tau_1(u), \ldots, \tau_k(u))$, where $\tau_i(u)$ lies in $U_i \backslash U_{i+1}$ and $v \in U_k$, then $\tau(uv) = (\tau_1(u), \ldots, \tau_{k-1}(u), \tau_k(u) + \tau(v))$. Taking this into account, we can replace the group U by $U \backslash U_k$, and correspondingly reduce V so that induction can be used.

In conclusion the author expresses thanks to O. V. Mel'nikov and A. S. Rapinchuk for useful discussions.

Bibliography

1. M. S. Raghunathan, *On the congruence subgroup problem*, Inst. Hautes Études Sci. Publ. Math. **46** (1976), 107–161.
2. Claude Chevalley, *Deux theoremes d'arithmetique*, J. Math. Soc. Japan **3** (1951), 36–44.
3. V. P. Platonov, *The congruence problem for solvable integral groups*, Dokl. Akad. Nauk BSSR **15** (1971), no. 6, 869–872. (Russian)
4. V. P. Platonov and A. A. Sharomet, *The congruence subgroup problem for linear groups over arithmetic rings*, Dokl. Akad. Nauk BSSR **16** (1972), 393–396. (Russian)
5. Andre Weil, *Basic number theory*, Springer-Verlag, New York, 1967.

Belorussian State University

Received 6/MAR/86

Translated by H. H. McFADEN

K-Theory of Free Products and the Bass Problem

UDC 512.733

A. V. PRASOLOV

Abstract. The paper contains a description of the K-groups of free products of rings with several objects, the partial solution of the problem of Bass, and the refinement of the Stallings theorem.

The goal of this paper is to describe the groups K_n of a free product of rings with amalgamated subring (Theorem 1), which enables us, in particular, to solve the famous Bass problem under certain restrictions (Theorem 3). In the case $n = 1$ it is possible to get rid of these restrictions (Theorem 2), i.e., to get an improvement of the Stallings result on the group K_1 of free products ([1], Chapter XII, Theorem 11.1). All the results in this article are formulated and proved in the language of pre-additive categories or of "rings with several objects" ([2], [3]), which form a natural generalization of the concept of a ring (associative with an identity). According to Mitchell, small pre-additive categories will be called *ringoids*. Of course, Theorems 1–3 remain true if in them the word "ringoid" is replaced by the word "ring".

Let C be a ringoid with set $|C|$ of objects, and S a C-C-bimodule. Denote by $NIL(C; S)$ the exact category (see [4] for the definition of an exact category and of K-functors of exact categories and rings) of ordered pairs (P, f), where $P \in \mathscr{P}_C$ ($\mathscr{P}_C$ is the category of finitely generated projective right C-modules), and $f: P \to \mathscr{P} \otimes_C S$ is a nilpotent C-homomorphism. Then $K_i NIL(C; S) \simeq K_i C \oplus$ is an Abelian group, which we denote by $Nil_i(C; S)$, for all $i \geq 0$.

PROPOSITION 1. *If a ringoid C is coherent and right regular, and if $S(-, j)$ is a flat left C-module for all $j \in |C|$, then $Nil_i(C; S) = 0$ for $i \geq 0$.*

1991 *Mathematics Subject Classification.* Primary 18F25, 19D35; Secondary 18E05.
Translation of Dokl. Akad. Nauk BSSR **25** (1981), no. 7, 598–600.

PROOF. Let $\mathcal{N} \supset NIL(C;S)$ be an Abelian category of ordered pairs (M, f), where M is a finitely representable right C-module, and $f: M \to M \otimes S$ is a nilpotent C-homomorphism. Then by "reduction by resolution" ([4], Theorem 3, p. 108), the imbedding $NIL(C;S) \to \mathcal{N}$ induces an isomorphism of the groups K_i, since any object in $\mathcal{N}$ admits a finite resolution from objects in the category $NIL(C;S)$. Moreover, by "reduction by unscrewing" ([4], Theorem 4, p. 112), the functor $P \to (P, 0)$ induces an isomorphism $K_i C \xrightarrow{\sim} K_i \mathscr{P}_C \xrightarrow{\sim} K_i \mathcal{N}$, since any object in $\mathcal{N}$ admits a finite filtration with factors of the form $(P, 0)$. Therefore, $K_i C \xrightarrow{\sim} K_i NIL$, $Nil_i(C;S) = 0$, as required.

CONJECTURE. Proposition 1 remains true if only the finiteness of the Tor-dimension of the left module S is required.

A *pure imbedding* of the ringoid C in the ringoid A is defined to be a morphism $f: C \to A$ such that $|f|: |C| \to |A|$ is a bijection, and $f(-,-): C(-,-) \to A(-,-)$ is a splitting monomorphism of C-C-bimodules (the cokernel of this monomorphism will be denoted by $A'(-,-)$).

Let $A \leftarrow C \to B$ be a diagram of pure imbeddings, and let $R = A *_C B$ be the direct limit of this diagram in the category of ringoids. Let $\mathscr{MV}'$ be the exact category of triples (M, N, f), where $M \in \mathscr{P}_A$, $N \in \mathscr{P}_B$, $f: M \otimes_A R \to N \otimes_B R$ is an R-homomorphism, and $f(M) \subset NA$. Further, let $\mathscr{MV} \subset \mathscr{MV}'$ be the full exact subcategory of (M, N, f) such that f is an epimorphism, and $\mathscr{V} \subset \mathscr{MV}$ is the full exact subcategory of triples (M, N, f) such that f is an isomorphism.

LEMMA 1. *For all i the functors $p_A: (M, N, f) \mapsto M$ and $p_B: (M, N, f) \mapsto N$ induce isomorphisms $K_i \mathscr{MV} \to K_i A \times K_i B$.*

PROOF. The imbedding $\mathscr{MV} \to \mathscr{MV}'$ satisfies the conditions of the theorem dual to the Quillen theorem on "reduction by resolution", therefore, $K_i(\mathscr{MV}) \to K_i(\mathscr{MV}')$ is an isomorphism. At the same time, the functors $s_A: P \mapsto (P, 0, 0)$ and $s_B: Q \mapsto (0, Q, 0)$ induce a homomorphism $K_i A \oplus K_i B \to K_i \mathscr{MV}'$ that is inverse to the homomorphism $K_i p_A \times K_i p_B$ in view of the short exact sequence of functors $s_B \circ p_B \rightarrowtail \text{Id} \twoheadrightarrow s_A \circ p_A$.

LEMMA 2. *There exists an exact equivalence of categories*

$$\mathscr{V} \cong NIL(C; B' \otimes_C A').$$

PROOF. If $(P, \varphi) \in NIL(C; B' \otimes A')$, then $1 + \varphi: P \otimes_C R \to P \otimes_C R$ is an isomorphism, therefore, the correspondence $(P, \varphi) \mapsto (P \otimes_C A, P \otimes_C B, 1 + \varphi)$ determines an exact functor $F: NIL(C; B' \otimes A') \mapsto \mathscr{V}$. But if $(M, N, f) \in \mathscr{V}$, then f takes the form $(M_1 \oplus (M_1 \otimes_C B') \to N \oplus N_1 \oplus (N_1 \otimes_C B'))$. If $K = \ker(M_1 \to N_1)$, then $N \simeq K \otimes_C B$, $M \simeq K \otimes_C A$, and a morphism $\varphi: K \to K \otimes B' \otimes A'$ such that $F(K, \varphi) = (M, N, f)$ is uniquely determined. The functor F thereby implements an exact equivalence $NIL(C: B' \otimes A') \to \mathscr{V}$.

THEOREM 1. *Suppose that for all $j \in |C|$ the left C-modules $A'(-, j)$ and $B'(-, j)$ are free. Then for $i \geq 1$ the decomposition $K_i R \simeq X_i \oplus Nil_{i-1}(C; B' \otimes A')$ holds, and X_i is included in the exact sequence*

$$\cdots \to K_i C \to K_i A \oplus K_i B \to X_i \to K_{i-1} C$$
$$\to \cdots \to X_1 \to K_0 C \to K_0 A \oplus K_0 B.$$

PROOF. An argument analogous to the proof of Proposition 10.1 in [5] proves the exactness of the sequence

$$\cdots \to K_i \mathcal{V} \to K_i \mathcal{M} \mathcal{V} \to K_i R \to K_{i-1} \mathcal{V} \to \cdots \to K_0 \mathcal{M} \mathcal{V},$$

and $K_i \mathcal{V} \simeq K_i C \oplus Nil_i(C; B' \otimes A')$, $K_i \mathcal{M} \mathcal{V} \simeq K_i A \oplus K_i B$ in view of Lemmas 1 and 2. This yields a decomposition of $K_i R$ into a direct sum, and the factorization of this exact sequence by $Nil_*(C, B' \otimes A')$ gives the desired exact sequence for X_i.

Suppose that the morphisms in the diagram $A \leftarrow C \to B$ are splitting monomorphisms of ringoids. Bass ([6], Problem VIII$_n$) formulated a conjecture (for rings): for all $i \geq 1$ the sequence

$$\overline{K}_i(T_C(B' \otimes A')) \to \overline{K}_i R \to \overline{K}_i A \oplus \overline{K}_i B \to 0$$

is exact, where $T_C(S)$ is the tensor algebra of the C-C-bimodule S, and $\overline{K}_i A = \operatorname{coker}(K_i C \to K_i A)$, and so on. We prove a stronger result for $i = 1$, and under the restrictions of Theorem 1 also for all $i \geq 1$.

THEOREM 2. *The sequence*

$$0 \to \overline{K}_1(T_C(B' \otimes A')) \to \overline{K}_1 R \to \overline{K}_1 A \oplus \overline{K}_1 B \to 0$$

is exact.

PROOF. We construct a mapping $\partial: K_1 R \to K_0 \mathcal{V}$ as follows: Let an element in $GL(R)$ be given by a matrix $x \in GL(\sigma, R)$, where σ is a finite $|R|$-set (i.e., equipped with a mapping $|\ |: \sigma \to |R|$). Then with the help of the "Higman trick" ([1], Proposition 5.1, p. 491) we reduce this matrix by elementary transformations and augmentations to a matrix $f \in GL(\tau, R)$ for which all terms belong to $C \oplus A' \oplus B' \subset R$. Then we set $\partial x = \partial f = [A^\tau, B^\tau, f] - [A^\tau, B^\tau, \operatorname{Id}] \in K_0 \mathcal{V}$, where $A^\tau = \coprod_{s \in \tau} A(|s|, -)$ is a free A-module.

LEMMA 3. *The mapping $\partial: K_1 R \to K_0 \mathcal{V}$ is well defined and is a homomorphism.*

PROOF. We represent all the terms of the matrix x as sums of products of elements in A or B. Then the matrix f is uniquely determined to within permutations of rows or columns. If some other representation as sums of products is possible, then it is possible to pass to it from the first representation by means of a finite collection of substitutions of the type

$$a\lambda \otimes b \leftrightarrow a \otimes \lambda b, \qquad (a_1 + a_2) \otimes b \leftrightarrow a_1 \otimes b + a_2 \otimes b,$$
$$b_1 \cdot b_2 \leftrightarrow (b_1 b_2), \qquad a_1 \cdot a_2 \leftrightarrow (a_1 a_2),$$

etc. With the help of the following lemma we see that the value of ∂x does not change under these substitutions, nor under elementary transformations.

LEMMA 4. *The following operations on f do not lead to a change of the element $\partial f = [A^\tau, B^\tau, f] - [A^\tau, B^\tau, \mathrm{Id}] \in K_0 \mathscr{V}$: a) the substitutions $f \leftrightarrow \bar{f} = \begin{pmatrix} f & * \\ 0 & 1 \end{pmatrix}$; b) left multiplications by elements in $GL(\tau, B)$; c) right multiplications by elements in $GL(\tau, A)$.*

PROOF. a) follows from the exact sequence
$$0 \to (A^{\tau_1}, B^{\tau_1}, f) \to (A^\tau, B^\tau, \tilde{f}) \to (A^{\tau_2}, B^{\tau_2}, \mathrm{Id}) \to 0,$$
and b) and c) lead to objects in $\mathscr{V}$ that are isomorphic to the original object.

Returning to the proof of the theorem, we show that the sequence
$$K_1 C \to K_1 A \oplus K_1 B \to K_1 R \xrightarrow{\partial} K_0 \mathscr{V} \xrightarrow{\Delta} K_0 A \oplus K_0 B$$
is exact, where $\Delta([P, Q, f]) = ([P], [Q]) \in K_0 A \oplus K_0 B$. The exactness at the terms $K_1 R$ and $K_0 \mathscr{V}$ can be proved by an argument analogous to Theorem 5.3 in Chapter IX of [1], with use of the isomorphism $K_0 \mathscr{V} \simeq K_0 C \oplus Nil_0(C; B' \otimes A')$, which follows from Lemma 2. This gives us the desired decompositions
$$K_1(T_C(B' \otimes A')) \simeq K_1 C \oplus Nil_0(C; B' \otimes A'),$$
$$\overline{K}_1 R \simeq \overline{K}_1 A \oplus \overline{K}_1 B \oplus Nil_0(C; B' \otimes A').$$

THEOREM 3. *If for all $j \in |C|$ the left C-modules $A'(-, j)$ and $B'(-, j)$ are free, then for $i \geq 1$ the sequence*
$$0 \to \overline{K}_i(T_C(B' \otimes A')) \to \overline{K}_i R \to \overline{K}_i A \oplus \overline{K}_i B \to 0$$
is exact.

PROOF. By Theorem 1, $\overline{K}_i R \simeq \overline{K}_i A \oplus \overline{K}_i B \oplus Nil_{i-1}(C; B' \otimes A')$. We consider the diagram of ringoids $C \leftarrow C \cup C \to D$, where D is the ringoid with two objects $\{i, j\}$ and with set of arrows $D(i, j) = S$, $D(j, i) = 0$, $D(i, i) = D(j, j) = C$. The direct limit of this diagram is isomorphic to $T_C(S)$, and $K_i D \simeq K_i C \oplus K_i C$ in view of Theorem 6 in [3], therefore, application of Theorem 1 to this diagram gives us that $K_i(T_C(S)) \simeq K_i C \oplus Nil_{i-1}(C; S)$, which concludes the proof of the theorem.

BIBLIOGRAPHY

1. Hyman Bass, *Algebraic K-theory*, Benjamin, New York, 1968.
2. Barry Mitchell, *Rings with several objects*, Adv. Math. **8** (1972), 1–161.
3. A. V. Prasolov, *Preadditive categories and K-theory*, Uspekhi Mat. Nauk **32** (1977), no. 5, 195–196. (Russian)
4. D. Quillen, *Higher algebraic K-theory. I*, Lecture Notes in Math., vol. 341, Springer-Verlag, Berlin and New York, 1973, pp. 85–147.

5. F. Waldhausen, *Algebraic K-theory of generalized free products.* I, II, Ann. of Math. (2) **108** (1978), no. 1, 135–204; III, IV, Ann. of Math. (2) **108** (1978), no. 2, 205–256.
6. Hyman Bass, *Some problems in "classical" algebraic K-theory*, Lecture Notes in Math., vol. 342, Springer-Verlag, Berlin and New York, 1973, pp. 3–73.

Belorussian State University

Received 26/NOV/80

Translated by H. H. McFADEN

On a Theorem of Gersten for Graded Rings

A. V. PRASOLOV

Gersten ([1], Theorem 5) computed the K-functors of a graded ring A (associative with an identity) in the case when $A[t]$ is a coherent regular ring. The goal of the present article is to compute the groups $K_i A$ for arbitrary A. This can be done with the help of the K-theory developed in [2] for pre-additive categories.

Let C be a small pre-additive category with set I of objects, and let $\alpha: C \to C$ be an automorphism.

DEFINITION. We define an α-twisted polynomial extension (Laurent extension) of the pre-additive category C to be a universally repelling object in the category of triples of the form (B, f, t), where B is a pre-additive category, $f: C \to B$ is an additive functor, and $t: f \circ \alpha \to f$ is a functorial morphism (isomorphism).

The category B defined in this way is denoted by $C_\alpha[t]$ in the case of a polynomial extension, and by $C_\alpha[t, t^{-1}]$ in the case of a polynomial Laurent extension. When the set I consists of a single element, i.e., when C is a ring, $C_\alpha[t]$ ($C_\alpha[t, t^{-1}]$) is the usual ring of α-twisted polynomials (Laurent polynomials).

Denote by $NIL(\alpha^{\pm 1}, C)$ the category of $\alpha^{\pm 1}$-linear nilpotent endomorphisms [3]. Let $Nil_i(\alpha^{\pm 1}, C) = \ker(K_i NIL(\alpha^{\pm 1}, C) \to K_i C)$. The K-functors of the ring of α-twisted Laurent polynomials are computed in [3]. This result is easy to generalize to pre-additive categories:

THEOREM 1. *For all* $i \geq 1$: a) $K_i C_{\alpha^{\pm 1}}[t] \simeq K_i C \dotplus Nil_{i-1}(\alpha^{\mp 1}, C)$;
b) $K_i C_\alpha[t, t^{-1}] \simeq X_i \dotplus Nil_{i-1}(\alpha, C) \dotplus Nil_{i-1}(\alpha^{-1}, C)$, *and* X_i *is included*

1991 *Mathematics Subject Classification.* Primary 19D25, 19D50.
Translation of Vestnik Beloruss. Gos. Univ. Ser. I **1981**, no. 2, 62–63.

in the exact sequence

$$K_iC \xrightarrow{\varphi_i} K_iC \to X_i \to K_{i-1}C \xrightarrow{\varphi_{i-1}} K_{i-1}C,$$

where $\varphi_* = \mathrm{Id} - K_*\alpha$.

Let $A = A_0 \dotplus A_1 \dotplus \cdots$ be a graded ring. Denote by $PGR(A)$ the category of finitely generated projective graded left A-modules, and by $NGR(A)$ the category of pairs of the form (P, f), where $P \in PGR(A)$, and f is a nilpotent endomorphism of degree 1. Let $Ngr_iA = \ker K_i(NGR(A) \to PGR(A))$.

THEOREM 2. *For all* $i \geq 1$

$$K_iA \simeq K_iA_0 \dotplus Ngr_{i-1}A.$$

PROOF. Let $C = gr(A)$ be the pre-additive category with set Z of objects, sets $C(m, n) = A_{m-n}$ of morphisms for $m \geq n$, and composition induced by the multiplication in the ring A. Then left C-modules are Z-graded left A-modules, and the morphisms of C-modules are graded homomorphisms of degree 0. Let $\alpha \colon C \to C$ be the shift automorphism $(n \to n+1)$. We have the additive equivalences of categories $C_\alpha[t, t^{-1}] \simeq A$ and $C_\alpha[t] \simeq gr(A[t])$, where $A[t]$ is the graded ring of polynomials in the variables t of degree 1. The category $NIL(\alpha, C)$ is equivalent to the category $NGR(A)$, therefore, the groups $Nil_i(\alpha, C)$ and Ngr_iA are isomorphic. Further, by [4] (p. 107), $K_i(gr(A[t])) \simeq K_i(gr(A)) \simeq K_iA_0 \otimes_Z Z[t, t^{-1}]$, therefore, $Nil_i(\alpha^{-1}, C) = 0$, $K_iA \simeq X_i \dotplus Ngr_{i-1}A$ by Theorem 1, and $X_i = \mathrm{coker}(\mathrm{Id} - \text{"shift"}) \simeq K_iA_0$, which implies the assertion of the theorem.

BIBLIOGRAPHY

1. S. Gersten, *K-theory of free rings*, Comm. Algebra **1** (1974), no. 1, 39.
2. A. V. Prasolov, *Preadditive categories and K-theory*, Uspekhi Mat. Nauk **32** (1977), no. 5, 195–196. (Russian)
3. ____, Fifteenth All-Union Conference Algebra, Part 1, Krasnoyarsk, 1979. (Russian)
4. D. Quillen, *Higher algebraic K-theory*. I, Lecture Notes in Math., vol. 341, Springer-Verlag, Berlin and New York, 1973, pp. 85–147.

Department of Higher Mathematics

Received 24/JAN/80

Translated by H. H. McFADEN

Algebraic K-Theory of Banach Algebras

A. V. PRASOLOV

ABSTRACT. For commutative Banach algebras A, functorial isomorphisms $K_i(A, \mathbf{Z}/m) \simeq K_i^{\text{top}}(A, \mathbf{Z}/m)$ are constructed between the algebraic and topological K-theories with coefficients $\mathbf{Z}/m$.

Everywhere in this paper A is a Banach algebra with identity over the field of real or complex numbers. Denote by $K_i(A, \mathbf{Z}/m) = \pi_i(BGL(A)^+, \mathbf{Z}/m)$ the Quillen K-functors ([1], [2]), and by $GL(A)$ the discrete group of invertible matrices over A. Adjoining the notation "top" $(GL(A)^{\text{top}}, A^{*\,\text{top}},$ etc.) will mean that the topology induced by the topology on A is considered on $GL(A), \ldots$, and $K_i^{\text{top}}(A), K_i^{\text{top}}(A, \mathbf{Z}/m)$ denote the homotopy groups of the space $BGL(A)^{\text{top}}$.

Our goal is to construct an isomorphism between the groups $K_i^{\text{top}}(A, \mathbf{Z}/m)$ and $K_i(A, \mathbf{Z}/m)$ for a commutative Banach algebra A. The proof of the main theorem is analogous in many respects to Suslin's proof in the case $A = \mathbf{R}$ or $\mathbf{C}$, in particular, it makes essential use of the "universal homotopy construction" due to Suslin.

We consider the natural continuous homomorphism $\varphi \colon GL(A) \to GL(A)^{\text{top}}$ and the mapping $B\varphi \colon BGL(A) \to BGL(A)^{\text{top}}$. Since $BGL(A)^{\text{top}}$ is an H-space, there exists a unique extension $B\varphi^+ \colon BGL(A)^+ \to BGL(A)^{\text{top}}$ of the mapping $B\varphi$.

THEOREM 1. *Let A be a commutative Banach algebra, and m an arbitrary positive integer. Then the mapping $B\varphi^+$ induces an isomorphism $K_*(A, \mathbf{Z}/m) \xrightarrow{\sim} K_*^{\text{top}}(A, \mathbf{Z}/m)$.*

We precede the proof of the theorem by some propositions and lemmas.

Let G be a topological group. Denote by BG^{top} (respectively, BG) the classifying space [3] of the topological group G (respectively, of the discrete

---

1991 *Mathematics Subject Classification.* Primary 46J99, 19K99, 19D99.
Translation of Dokl. Akad. Nauk BSSR **28** (1984), no. 8, 677–679.

group G). Let U be an open neighborhood of the identity in G, and let $BG_U \subset BG$ consist of simplices $(g_1, g_2, \ldots, g_n) \in (BG)_n$ for which $U \cap g_1 U \cap \cdots \cap g_1 g_2 \cdots g_n U \neq \varnothing$. We say that the neighborhood U is marked if for any $g_0, \ldots, g_n \in G$ the space $g_0 U \cap g_1 U \cap \cdots \cap g_n U$ is contractible or empty.

LEMMA 1. *If U is a marked neighborhood, then*
$$BG_U \to BG \to BG^{\text{top}}$$
is a fibration up to homotopy.

Let A be a commutative Banach algebra. Then denote by $U(n, \varepsilon)$ the set of $n \times n$ matrices over A of the form $1 + h$, where $\rho(h_{ij}) < \varepsilon$ for all $1 \leq i, j \leq n$ (here $\rho(h) = \inf\{\|h^n\|^{1/n}\}$ is the spectral radius of an element h).

LEMMA 2. *If $0 < \varepsilon \leq 1/n$, then $U(n, \varepsilon)$ is a marked neighborhood of the identity of the topological group $GL_n(A)$.*

Denote by $\mathscr{O}_{n,i}$ and $\mathscr{M}_{n,i}$ the local ring of the real analytic manifold $GL_n \times \cdots \times GL_n$ (i times) at the point $(1, \ldots, 1)$ and its maximal ideal. The boundary morphisms $d^i_j: (GL_n)^i \to (GL_n)^{i-1}$ of the symplicial manifold BGL_n determine ring homomorphisms $p^i_j: \mathscr{O}_{n,i-1} \to \mathscr{O}_{n,i}$ ($j = 0, 1, \ldots, i$). Let $C_*(G, \mathbf{Z}/m)$ be the reduced chain complex of the space BG. The germs of the coordinate functions x^j_{kl}, $1 \leq k, l \leq n$, $1 \leq j \leq i$, determine the element $u_{n,i} = ((x^1_{kl}), (x^2_{kl}), \ldots, (x^i_{kl})) \in C_i(GL(\mathscr{O}_{n,i}; \mathscr{M}_{n,i}), \mathbf{Z}/m)$ (we set $u_{n,0} = 0$).

LEMMA 3 (universal homotopy construction). *There exist chains $c_{n,i} \in C_{i+1}(GL(\mathscr{O}_{n,i}; \mathscr{M}_{n,i}), \mathbf{Z}/m)$ such that*
$$d(c_{n,i}) = u_{n,i} - \sum_{j=0}^{i}(p^i_j)^*(c_{n,i-1}),$$
where d is the differential in the chain complex $C_*(GL(\mathscr{O}_{n,i}; \mathscr{M}_{n,i}), \mathbf{Z}/m)$.

PROPOSITION 1. *Let A be a commutative Banach algebra, and F the homotopy fiber of the mapping $B\varphi: BGL(A) \to BGL(A)^{\text{top}}$. Then $H_i(F, \mathbf{Z}/m) = 0$ for $i \geq 1$.*

PROPOSITION 2. *Under the conditions of Proposition 1 let $\psi_m: A^* \to A^*$ be the raising to the power m. Then ψ_m is a fibering in the sense of Serre, and the fiber X of this fibering satisfies the condition $\pi_0 X = X$.*

The following theorem can be proved by an argument analogous to the proof of Theorem 1.

THEOREM 2. *Suppose that A is a not necessarily commutative Banach algebra, $q \subset \operatorname{Rad} A$ is a closed ideal, and $p: A \to A/q$ is the projection. Then*: a) $K_0 p$ *is an isomorphism*; b) *if A is commutative, then $K_i(p, \mathbf{Z}/m\mathbf{Z})$ is an isomorphism for all $i \geq 1$*.

PROOF OF THEOREM 1. By Proposition 2, the group $\pi_i(BA^{*\,\mathrm{top}}, \mathbf{Z}/m\mathbf{Z})$ is equal to zero for $i \geq 3$, is isomorphic to $\pi_0 X \simeq X$ for $i = 2$, and is isomorphic to $\pi_0 A^*/(\pi_0 A^*) \simeq A^*/(A^*)^m$ for $i = 1$. Therefore, $\pi_i(BA^*, \mathbf{Z}/m) \to \pi_i(BA^{*\,\mathrm{top}}, \mathbf{Z}/m)$ is an isomorphism for all $i \geq 1$. An analogous assertion is true also for numbers of the form m^k, and since BA^* and $BA^{*\,\mathrm{top}}$ are nilpotent spaces [4], it follows that $(\mathbf{Z}/m)_\infty BA^* \to (\mathbf{Z}/m)_\infty BA^{*\,\mathrm{top}}$ is a homotopy equivalence. Thus, $H^*(BA^*, \mathbf{Z}/m) \to H^*(BA^{*\,\mathrm{top}}, \mathbf{Z}/m)$ is an isomorphism.

Consider the commutative diagram of fiberings:

$$\begin{array}{ccccc} BSL(A)^+ & \longrightarrow & BGL(A)^+ & \longrightarrow & BA^* \\ \varphi_1 \downarrow & & \varphi_2 \downarrow & & \varphi_3 \downarrow \\ BSL(A)^{\mathrm{top}} & \longrightarrow & BGL(A)^{\mathrm{top}} & \longrightarrow & BA^{*\,\mathrm{top}} \end{array}$$

It was just proved that $H_*(\varphi_3, \mathbf{Z}/m)$ is an isomorphism, and $H_*(\varphi_2, \mathbf{Z}/m)$ is an isomorphism in view of Proposition 1. According to the comparison theorem for spectral sequences, $H_*(\varphi_1, \mathbf{Z}/m)$ is also an isomorphism. But $\pi_1 \varphi_1$ is an isomorphism (see [5], §7), therefore, $\pi_i(\varphi_1, \mathbf{Z}/m)$ is an isomorphism for all $i \geq 1$. By virtue of the five-lemma, $\pi_i(\varphi_2, \mathbf{Z}/m)$ is an isomorphism for $i \geq 1$.

PROOF OF PROPOSITION 1. Denote by F_n the homotopy fiber of the mapping $BGL_n(A) \to BGL_n(A)^{\mathrm{top}}$. Since $F = \lim_n F_n$, and $F_n \simeq U(n, \varepsilon)$ for $\varepsilon \leq 1/n$ in view of Lemmas 1 and 2, it suffices to prove the following assertion: for any n there is an N such that for arbitrary $\varepsilon > 0$ and i_0 there exists a δ with $\varepsilon > \delta > 0$ for which the imbedding $U(n, \delta) \to U(N, \varepsilon)$ induces the zero homomorphism

$$H_i(BGL_n(A)_\delta, \mathbf{Z}/m) \to H_i(BGL_N(A)_\varepsilon, \mathbf{Z}/m),$$

$1 \leq i \leq i_0$. To prove this last assertion we use Lemma 3. Let $a = ((a_{kl}^1), (a_{kl}^2), \ldots, (a_{kl}^i)) \in C_i(BGL_n(A)_\delta, \mathbf{Z}/m)$. The germs of the functions appearing in the elements $c_{n,i}$ for $i \leq i_0$ are defined on some neighborhood of the point $(1, \ldots, 1)$, therefore, for sufficiently small δ the elements a_{kl}^j can be substituted in these functions in place of the variables x_{kl}^j ([6], Chapter 1, §4), which enables us to construct the elements $c_{n,i}(a) \in C_{i+1}(BGL_N(A), \mathbf{Z}/m)$. By decreasing δ we can ensure that $c_{n,i}(a) \in C_{i+1}(BGL_N(A)_\varepsilon, \mathbf{Z}/m)$. The mapping $a \to c_{n,i}(a)$ determines the desired chain homotopy $C_i(BGL_n(A)_\delta, \mathbf{Z}/m) \to C_{i+1}(BGL_N(A)_\varepsilon, \mathbf{Z}/m)$.

PROOF OF PROPOSITION 2. Note that $\psi_m(A^*) = (A^*)^m$ is an open (and thus also closed) subgroup of A^*. Let $y \in \psi_m(A)$, i.e., $y = x_0^m$, and let

$U = \{z \in A^* | \rho(1-z) < 1\}$. Then yU is an open neighborhood of y, and it suffices to show that $\psi_m^{-1}(yU) \to yU$ is a fiber space in the Serre sense. Let $X_y = \{x \in A^* | x^m = y\}$, $x \in X_y$. We consider the mapping $\xi_x : yU \to A^*$ given by the formula

$$\xi_x(y(1+h)) + x \cdot \sum_{i=0}^{\infty} \binom{1/n}{i} h^i.$$

It is easy to see that $\psi_m \circ \xi_x = \mathrm{Id}$, therefore, $\xi_x : yU \to \xi_x(yU)$ is a homeomorphism, and

$$\psi_m^{-1}(yU) = \bigcup_{x \in X_y} \xi_x(yU).$$

It remains to show that no two distinct points of the set X_y can be joined by a continuous path lying in $\psi_m^{-1}(yU)$. But if the points $x_1, x_2 \in X$ can be joined by such a path, then the spectrum of the element $x_1 x_2^{-1}$ is equal to $\{1\}$, therefore, $x_1 x_2^{-1} - 1 \in \mathrm{Rad}\, A$, but this is impossible by virtue of the following lemma.

LEMMA 4. *If $q \subset \mathrm{Rad}\, A$ is a closed ideal, then the raising to the mth power is an automorphism of the topological group $1 + q \subset A^*$.*

PROOF OF THEOREM 2. a) The injectivity of $K_0 p$ and the surjectivity of $K_1 p$ were proved in [7] (pp. 448–449). There also it was proved that $\ker K_1 p = 1 + q$ if A is commutative. The matrix algebra $M_n A$ admits the structure of a Banach algebra, and $M_n q \subset \mathrm{Rad}\, M_N A$, therefore, the idempotents in the algebra $M_n(A/q)$ can be lifted modulo the ideal $M_n q$, and the homomorphism $K_0 p$ is surjective. b) If A is commutative, then $\ker K_1 p = 1 + q$ is a uniquely m-divisible group in view of Lemma 4, therefore, it suffices to show that $H_i(GL(A, q), \mathbf{Z}/m) = 0$ for $i \geq 1$. Returning to the proof of Proposition 1, we note that if $a_{kl}^j \in GL_n(A, q)$, then $c_{n,i}(a) \in C_{i+1}(GL_N(A, q), \mathbf{Z}/m)$, therefore, the correspondence $a \to c_{n,i}(a)$ determines a chain homotopy between the imbedding $C_*(GL_n(A, q), \mathbf{Z}/m) \to C_*(GL_N(A, q), \mathbf{Z}/m)$ and the zero mapping. Thus, $H_i(GL(A, q), \mathbf{Z}/m) = \lim H_i(GL_n(A, q), \mathbf{Z}/m) = 0$, which is what was required.

BIBLIOGRAPHY

1. A. A. Suslin, *Algebraic K-theory*, Itogi Nauki i Tekhniki: Algebra, Topologiya, Geometriya, vol. 20, VINITI, Moscow, 1982, pp. 71–152; English transl. in J. Soviet Math. **28** (1985), no. 6.
2. W. Browder, *Algebraic K-theory with coefficients $\widetilde{Z}/p$*, Lecture Notes in Math., vol. 657, Springer-Verlag, Berlin and New York, 1978, pp. 40–84.
3. Graeme Segal, *Classifying spaces and spectral sequences*, Inst. Hautes Études Sci. Publ. Math. No. 34, 1968, pp. 105–112.
4. A. K. Bousfield and D. M. Kan, *Homotopy limits, completions and localizations*, Lecture Notes in Math., vol. 304, Springer-Verlag, Berlin and New York, 1972.

5. J. Milnor, *Introduction to algebraic K-theory*, Princeton Univ. Press, Princeton, NJ, 1971.
6. N. Bourbaki, *Théorie spectrales*, Chaps. 1, 2, Actualités Sci. Indust., no. 1332, Hermann, Paris, 1967.
7. Hyman Bass, *Algebraic K-theory*, Benjamin, New York, 1968.

Belorussian State University

Received 14/OCT/83

Translated by H. H. McFADEN

On the Finite Generation of the Character Ring of Three-Dimensional Unimodular Group Representations

UDC 512.547+512.552

V. V. BENYASH-KRIVETS

Abstract. It is proved that the ring of characters affording three-dimensional unimodular representations of a finitely generated group is finitely generated.

Let G be an arbitrary finitely generated group with generators $g_1, \ldots, g_n$. For an arbitrary field K, consider the set $V(G, SL_3(K))$ of all linear representations $\rho\colon G \to SL_3(K)$, where $SL_3(K)$ is the group of unimodular matrices of degree 3. The set $V(G, SL_3(K))$ can be canonically identified with an algebraic subset of $SL_3(K) \times \cdots \times SL_3(K)$ (n times). For each $g \in G$ we introduce a regular function $t(g)$ on $V(G, SL_3(K))$ by letting $t(g)\colon \rho \mapsto \operatorname{tr}\rho(g)$, where $\operatorname{tr}\rho(g)$ is the trace of the matrix $\rho(g)$. Let $T(G, SL_3(K))$ denote the ring generated by all functions $t(g)$, $g \in G$. This ring is called the character ring of three-dimensional unimodular representations of G, or briefly the character ring of G. Let us make some remarks on the functor properties of character rings. To any homomorphism of groups $\varphi\colon G \to G'$, there corresponds a homomorphism of the character rings $\overline{\varphi}\colon T(G, SL_3(K)) \to T(G', SL_3(K))$, $\overline{\varphi}(t(g)) = t(\varphi(g))$. If φ is an epimorphism, then $\overline{\varphi}$ is an epimorphism as well. On the other hand, if φ is injective, then $\overline{\varphi}$ need not be injective. Furthermore, if $L \supset K$ is a field extension, then $V(G, SL_3(K)) \subset V(G, SL_3(L))$ and therefore the functions from $T(G, SL_3(K))$ are just the restrictions to $V(G, SL_3(K))$ of the corresponding functions from $T(G, SL_3(L))$. Fricke and Klein [1] introduced the character ring $T(G, SL_2(K))$ of two-dimensional unimodular representations of G and conjectured that this ring is finitely generated. The ring $T(G, SL_2(K))$ is called the Fricke character ring. In 1972, Horowitz [2]

1991 *Mathematics Subject Classification.* Primary 20C99.
Translation of Dokl. Akad. Nauk BSSR **30** (1986), no. 5, 397–399.

©1992 American Mathematical Society
0065-9290/92 $1.00 + $.25 per page

proved that the functions $t(g_{i_1}, g_{i_2}, \ldots, g_{i_k})$, $1 \le i_1 < i_2 < \cdots < i_k \le n$, generate $T(G, SL_2(K))$. The main results concerning the Fricke character rings and their group-theoretic applications may be found in Magnus' survey [3].

In the present note we prove the following

THEOREM. *The character ring* $T(G, SL_3(K))$ *of three-dimensional unimodular representations of a finitely generated group* G *is finitely generated.*

The following lemma plays the key role in the proof.

LEMMA. *For arbitrary* $g, h \in G$

$$t(g^2 h) = t(g) \cdot t(gh) - t(g^{-1}) \cdot t(h) + t(g^{-1}h). \tag{1}$$

PROOF OF THE LEMMA. Take an arbitrary matrix $A \in SL_3(K))$, and let $\alpha_1, \alpha_2, \alpha_3$ be the roots of its characteristic polynomial. By the Cayley-Hamilton Theorem, A satisfies the equation

$$A^3 - A^2 \operatorname{tr} A + A(\alpha_1\alpha_2 + \alpha_1\alpha_3 + \alpha_2\alpha_3) = E = 0.$$

We note that $\operatorname{tr} A^{-1} = \alpha_1^{-1} + \alpha_2^{-1} + \alpha_3^{-1} = \alpha_1\alpha_2 + \alpha_1\alpha_3 + \alpha_2\alpha_3$ because $\alpha_1\alpha_2\alpha_3 = 1$. Therefore the above equation can be rewritten in the form

$$A^3 - A^2 \operatorname{tr} A + A \operatorname{tr} A^{-1} - E = 0. \tag{2}$$

Multiply both parts of (2) by $A^{-1}B$ on the right, where $B \in SL_3(K)$, and take the trace:

$$\operatorname{tr} A^2 B = \operatorname{tr} A \cdot \operatorname{tr} AB - \operatorname{tr} A^{-1} \cdot \operatorname{tr} B + \operatorname{tr} A^{-1} B. \tag{3}$$

This equality is valid for all $A, B \in SL_3(K)$, whence the lemma follows.

PROOF OF THE THEOREM. Since a finitely generated group is an epimorphic image of a free group, it follows from the preceding remarks that we can assume G to be free. We will show that there exists a positive integer N, depending on the number of generators $g_1, \ldots, g_n$ of the group G, such that for every word $W \in G$ the function $t(W)$ can be expressed as a polynomial over $\mathbf{Z}$ in $t(W_1), \ldots, t(W_k)$, where each W_i is of the form

$$W_i = g_{i_1}^{\varepsilon_1} \cdots g_{i_s}^{\varepsilon_s}, \qquad \varepsilon_j \in \{-1, 1\}, \ 1 \le j \le s,$$

and each generator g_r occurs at most N times in W_i. Since G is finitely generated, there is only a finite number of such words W_i, and this will immediately imply that the ring $T(G, SL_3(K))$ is finitely generated. We note that, in addition to (1), there is an evident equality $t(gh) = t(hg)$ for any $g, h \in G$, because $\operatorname{tr} AB = \operatorname{tr} BA$ for arbitrary matrices A and B.

Step 1. Consider an arbitrary $W = g_{i_1}^{n_1} \cdots g_{i_r}^{n_r}$ from G. Applying our Lemma a sufficient number of times, we can present $t(W)$ as a polynomial over **Z** in $T(W_1), \ldots, t(W_s)$, each W_i being of the form

$$W_i = g_{p_1}^{\varepsilon_1} \cdots g_{p_t}^{\varepsilon_t}, \tag{4}$$

where $\varepsilon_i \in \{-1, 1\}$, $1 \leq i \leq t$, and $p_1, \ldots, p_t$ need not be distinct.

Step 2. Let W have the form (4) and let the generator g_1 occur in W as follows:

$$W = W_1 g_1 W_2 g_1 W_3, \tag{5}$$

where W_2 does not contain g_1 (the case $W = W_1 g_1^{-1} W_2 g_1^{-1} W_3$ is treated in a similar way). Then

$$t(W) = t(W_1 g_1 W_2 g_1 W_3) = t(g_1 W_2 g_1 W_3 W_1)$$

(denote for brevity $W_2 = V_1$, $W_3 W_1 = V_2$)

$$= t(g_1 V_1 g_1 V_2) = t((g_1 V_1)^2 V^{-1} V_2)$$
$$= t(g_1 V_1 g_1 V_2) = t((g_1 V_1)^2 V_1^{-1} V_2) \quad \text{(by the Lemma)}$$
$$= t(g_1 V_1) t(g_1 V_2) - t((g_1 V_1)^{-1}) \cdot t(V_1^{-1} V_2) + t((g_1 V_1)^{-1} V_1^{-1} V_2).$$

Thus

$$\begin{aligned} t(g_1 V_1 g_1 V_2) &= t(g_1 V_1) t(g_1 V_2) - t((g_1 V_1)^{-1}) \cdot t(V_1^{-1} V_2) \\ &\quad + t(g_1^{-1} V_1^{-1} V_2 V_1^{-1}). \end{aligned} \tag{6}$$

If necessary, apply Step 1 to the functions $t(V_1^{-1} V_2)$ and $t(g_1^{-1} V_1^{-1} V_2 V_1^{-1})$. We obtain that if W is of the form (4) and satisfies (5), then $t(W)$ is a **Z**-polynomial in $t(W_1), \ldots, t(W_s)$, where each W_i has the form (4) as well, but the number of occurrences of g_1 in W_i is strictly less than in W. Applying (6) a sufficient number of times, we can express $t(W)$ as a polynomial over **Z** in the functions $t(W_i)$, where

$$W_i = g_1 V_1 g_1^{-1} V_2 g_1 \cdots g_1^{-1} V_s, \tag{7}$$

and g_1 is not involved in the V_j.

Step 3. Let W have the form (7), and suppose that the number of occurrences of g_1 in W is at least 4, i.e., $W = g_1 W_1 g_1^{-1} W_2$, where W_1 does not contain g_1 and W_2, by (7), has the form

$$W_2 = U_1 g_1 U_2 g_1^{-1} U_3, \tag{8}$$

where g_1 does not occur in U_1 and U_3. Then

$$\begin{aligned} t(W) &= t(g_1 W_1 g_1^{-1} W_2) = t(g_1^2 W_1 g_1^{-1} W_2 g_1^{-1}) \quad \text{(by the Lemma)} \\ &= t(g_1) \cdot t(W_1 g_1^{-1} W_2) - t(g_1^{-1}) \cdot t(W_1 g_1^{-1} W_2 g_1^{-1}) \\ &\quad + t(g_1^{-1} W_1 g_1^{-1} W_2 g_1^{-1}). \end{aligned} \tag{9}$$

Apply Step 2 to the function $t(W_1 g_1^{-1} W_2 g_1^{-1})$ from (9). Then consider the function $t(V) = t(g_1^{-1} W_1 g_1^{-1} W_2 g_1^{-1})$ from (9). According to (6),

$$\begin{aligned} t(V) &= t(g_1^{-1} W_1) \cdot t(g_1^{-1} W_2 g_1^{-1}) - t((g_1^{-1} W_1)^{-1}) \cdot t(W_2 g_1^{-1} W_1) \\ &\quad + t(g_1 W_1^{-1} W_2 g_1^{-1} W_1^{-1}) \\ &= t(g_1^{-1} W_1) \cdot t(g_1^{-2} W_2) - t(g_1 W_1^{-1}) \cdot t(W_2 g_1^{-1} W_1^{-1}) \\ &\quad + t(g_1 W_1^{-1} W_2 g_1^{-1} W_1^{-1}). \end{aligned} \tag{10}$$

Apply Step 1 to the function $t(g_1^{-2} W_2)$; then it remains to consider the function $t(g_1 W_1^{-1} W_2 g_1^{-1} W_1^{-1})$ from (10). Substituting in it (8), we obtain

$$\begin{aligned} t(g_1 W_1^{-1} W_2 g_1^{-1} W_1^{-1}) &= t(g_1 (W_1^{-1} U_1) g_1 (U_2 g_1^{-1} U_3 g_1^{-1} W_1^{-1})) \\ &= t(g_1 X_1 g_1 X_2), \end{aligned}$$

where $X_1 = W_1^{-1} U_1$ and $X_2 = U_2 g_1^{-1} U_3 g_1^{-1} W_1^{-1}$. Now we can apply (6) to $t(g_1 X_1 g_1 X_2)$ and then in turn Step 1 and, possibly, Step 2. In the long run, we will present $t(W)$ as a polynomial over $\mathbf{Z}$ in $t(V_1), \ldots, t(V_k)$, where each V_i is of the form (7) and the number of occurrences of g_1 in V_i is strictly less than that in W. Repeating this process, we eventually come to the situation when each V_i has the form

$$V_i = \begin{cases} W_1 g_1, \\ W_1 g_1^{-1}, \\ W_1 g_1^{-1} W_2 g_1, \end{cases} \tag{11}$$

where g_1 is involved neither in W_1, nor in W_2.

Step 4. Let W have the form (11), say $W = V_1 g_1^{-1} V_2 g_1$ (the cases $W = V_1 g_1$ or $W = V_1 g_1^{-1}$ are similar). It is now natural to repeat Steps 2 and 3 for the generator g_2. However, one should keep in mind that if we express $t(W)$ as a polynomial in $t(U_1), \ldots, t(U_k)$, where the number of occurrences of g_2 in the U_i is less than that in W, we must be sure that the number of occurrences of g_1 in the U_i is still at most two. In other words, while reducing the number of occurrences of g_2, we must be careful that the number of occurrences of g_1 is not increased. Note that the latter can become greater only when we apply (6). Let $W = W_1 g_2 W_2 g_2 W_3$. If g_1 is not involved in W_2, then any application of (6) does not increase the number of occurrences of g_1. Now we can proceed as follows. Applying Steps 1–3, we first reduce the number of occurrences of g_2 in V_1, and only after that in V_2 (in view of $t(V_1 g_1^{-1} V_2 g_1) = t(V_2 g_1 V_1 g_1^{-1})$). Eventually $t(W)$ will be expressed as a polynomial in $t(W_1), \ldots, t(W_k)$, where for each i the number of occurrences of g_1 in W_i is at most two, but that of g_2 is at most four. Let us express W_i in the explicit form $W_i = (V_1 g_2^{\varepsilon_1} V_2 g_2^{\varepsilon_2} V_3) g_1^{\varepsilon_4} (V_4 g_2^{\varepsilon_5} V_5 g_2^{\varepsilon_6} V_6)$, where $\varepsilon_i \in \{-1, 0, 1\}$, $1 \le i \le 6$. Repeating the above arguments, we will come to the situation when the number of occurrences of g_3 in each V_i,

$1 \leq i \leq 6$, is at most two. Together this yields at most 12 occurrences of g_3. Similarly, for g_k we will obtain at most $4 \cdot 3^{k-2}$ occurrences. It follows that for the desired number N one can take $4 \cdot 3^{n-2}$, where n is the number of generators of G. This completes the proof.

In conclusion, the author would like to express his gratitude to V. P. Platonov for numerous discussions and suggestions.

Bibliography

1. R. Fricke and F. Klein, *Vorlesungen über die Theorie der automorphen Funktionen*, Band 1, Teubner, Leipzig, 1897; Reprint, Johnson Reprint Corp., New York, 1965.
2. R. Horowitz, *Characters of free groups represented in the two-dimensional special linear group*, Comm. Pure Appl. Math. **25** (1972), no. 6, 635–649.
3. W. Magnus, *The uses of 2 by 2 matrices in combinatorial group theory. A survey*, Resultate Math. **4** (1981), no. 2, 171–192.

Institute of Mathematics
Academy of Sciences of the Belorussian SSR

Received 11/JUNE/85

Translated by S. M. VOVSI

The K_G-Functor on the Category of Inverse Spectra of Topological Spaces

UDC 513.836

S. A. BAĬRAMOV

Abstract. In this paper the K_G-functor on categories of inverse spectra of topological spaces is introduced and some properties of this functor are investigated.

The K-functor was constructed on the category of Boolean algebras with closure in [3]. Following the ideas in [3], we introduce here the category of vector G-bundles over an inverse spectrum of topological G-spaces. Then we define the K_G-functor and construct an extraordinary cohomology theory for inverse spectra of topological G-spaces. For each continuous G-mapping $f\colon X \to Y$ of topological G-spaces, denote by f^* the covariant functor induced by f, which associates with each G-bundle $p \in \operatorname{Vect}_G(Y)$ the induced bundle $f^*(p) \in \operatorname{Vect}_G(X)$.

The rule $X \to \operatorname{Vect}_G(X)$, $f \to f^*$ is a contravariant functor from the category Top_G of topological G-spaces to the category of categories. Consider the category $\operatorname{Invspec}(\operatorname{Top}_G)$ of inverse spectra of topological G-spaces. With each inverse spectrum

$$X \in \operatorname{Invspec}(\operatorname{Top}_G), \qquad X = (\{X_i\}_{i \in I},\ \{p_i^{i'}\colon X_{i'} \to X_i\}_{i < i' \in I})$$

of topological G-spaces we associate the direct spectrum

$$(\{\operatorname{Vect}_G(X_i)\}_{i \in I},\ \{(p_i^{i'})^*\colon \operatorname{Vect}_G(X_i) \to \operatorname{Vect}_G(X_{i'})\}_{i < i' \in I})$$

of categories of vector G-bundles, and with each morphism $f\colon X \to Y$ of the inverse spectrum $X = (\{X_i\}_{i \in I},\ \{p_i^{i'}\}_{i < i' \in I})$ to the inverse spectrum $Y = (\{Y_j\}_{j \in J},\ \{p_j^{j'}\}_{j < j' \in J})$, $f = (\pi\colon J \to I,\ \{f_j\colon X_{\pi(j)} \to Y_j\}_{j \in J})$, we associate

---

1991 *Mathematics Subject Classification.* Primary 55N20, 55N15.
Translation of Izv. Akad. Nauk Azerbaĭdzhan. SSR Ser. Fiz.-Tekhn. Mat. Nauk **1979**, no. 2, 3–9.

the morphism $f^*: \{\text{Vect}_G(Y_j)\}_{j \in J} \to \{\text{Vect}_G(X_i)\}_{i \in I}$,
$$f^* = (\pi: J \to I, \{f_j^*: \text{Vect}_G(Y_j) \to \text{Vect}_G(X_{\pi(j)})\}_{j \in J}).$$

The morphism f^* is well defined. Indeed, for any $j < j' \in J$ the commutativity of the diagram

$$\begin{array}{ccc} X_{\pi(j')} & \xrightarrow{f_{j'}} & Y_{j'} \\ \downarrow p_{\pi(j)}^{\pi(j')} & & \downarrow p_j^{j'} \\ X_{\pi(j)} & \xrightarrow{f_j} & Y_j \end{array}$$

implies the commutativity of the diagram

$$\begin{array}{ccc} \text{Vect}_G(X_{\pi(j')}) & \xrightarrow{f_{j'}^*} & \text{Vect}_G(Y_{j'}) \\ \uparrow (p_{\pi(j)}^{\pi(j')})^* & & \uparrow (p_j^{j'})^* \\ \text{Vect}_G(X_{\pi(j)}) & \xrightarrow{f_j^*} & \text{Vect}_G(Y_j) \end{array}$$

to within an isomorphism. The relation $X \to \{\text{Vect}_G(X_i)\}_{i \in I}$, $f \to f^*$ is a contravariant functor from the category $\text{Invspec}(\text{Top}_G)$ of inverse spectra of topological G-spaces to the category of direct spectra of categories of vector G-bundles.

Let $f = (\pi, \{f_j\}_{j \in J}): (\{X_i\}_{i \in I}, \{p_i^{i'}\}_{i < i' \in I}) \to (\{Y_j\}_{j \in J}, \{p_j^{j'}\}_{j < j' \in J})$ and $g = (\rho, \{g_k\}_{k \in K}): (\{Y_j\}_{j \in J}, \{p_j^{j'}\}_{j < j' \in J}) \to (\{Z_k\}_{k \in K}, \{p_k^{k'}\}_{k < k' \in K})$ be morphisms of inverse spectra, and $g \cdot f$ their composition. For any $k \in K$ the relation $(g_k \cdot f_{\rho(k)})^* = f_{\rho(k)}^* \cdot g_k^*$ holds, and then also $(g \cdot f)^* = f^* \cdot g^*$. If $(\text{id}, 1_X)$ is the identity morphism of the inverse spectrum X, then $(\text{id}, 1_x)^* = (\text{id}, \{1_{\text{Vect}_G(x_i)}\}_{i \in I}$. This proves our assertion.

DEFINITION 1. The category $\text{Vect}_G[X] = \varinjlim_{i \in I} \text{Vect}_G(X_i)$ is called the category of vector G-bundles over the inverse spectrum X. It is clear that the rule assigning to each inverse spectrum X of topological G-spaces the category $\text{Vect}_G[X]$ of vector G-bundles over the inverse spectrum X is a contravariant functor acting from the category $\text{Invspec}(\text{Top}_G)$ to the category of categories of vector G-bundles.

The functor $f^*: \text{Vect}_G(Y) \to \text{Vect}_G(X)$ of the induced bundle preserves Whitney sums of vector G-bundles: $f^*(p \circ q) = f^*(p) \circ f^*(q)$ for any vector G-bundles $p: E \to Y$ and $q: F \to Y$. Thus, for each continuous G-mapping $f: X \to Y$ the induced covariant functor $f^*: \text{Vect}_G(Y) \to \text{Vect}_G(X)$ is a morphism of semigroups. Therefore, the vector G-bundles $\text{Vect}_G[X]$ over an inverse spectrum X form a semigroup with respect to the operation $\circ$.

DEFINITION 2. The group $K_G[X] = K_G^n(\text{Vect}_G[X])$ that is the completion of the semigroup $\text{Vect}_G[X]$ of vector G-bundles over an inverse spectrum X of topological G-spaces is called the Grothendieck group of the inverse spectrum X.

The tensor product of two vector G-bundles over an inverse spectrum will be a vector G-bundle if it is assumed that G acts diagonally on the product. Consequently, under this condition $K_G[X]$ will be a commutative ring with identity.

Since the functor Vect_G acting from the category $\text{Invspec}(\text{Top}_G)$ of inverse spectra of topological G-spaces to the category of semirings is contravariant, while the completion functor from the category of semirings to the category of rings is covariant, their composition $K_G^n \circ \text{Vect}_G$ is a contravariant functor acting from the category $\text{Invspec}(\text{Top}_G)$ to the category Ring. Accordingly, we have

THEOREM 1. *The rule assigning to each inverse spectrum X of topological G-spaces the Grothendieck ring $K_G[X]$ of X and to each continuous G-morphism $f: X \to Y$ of inverse spectra the homomorphism $K_G[f]: K_G[Y] \to K_G[X]$ induced by the morphism $f_*^*: \text{Vect}_G[Y] \to \text{Vect}_G[X]$ of semirings, is a contravariant functor from the category* $\text{Invspec}(\text{Top}_G)$ *of inverse spectra of topological G-spaces to the category* Ring *of rings.*

For an inverse spectrum $X = (\{X_i\}_{i \in I}, \{p_i^{i'}: X_{i'} \to X_i\}_{i < i' \in I})$ of topological G-spaces let $K_G(X_i)$ be the Grothendieck ring of the topological G-space X_i for any $i \in I$. Then we have

THEOREM 2. $(\{K_G(X_i)\}_{i \in I}, \{K_G(p_i^{i'}): K_G(X_i) \to K_G(X_{i'})\}_{i < i' \in I})$ *is a direct spectrum of rings, and the limit $\varinjlim_{i \in I} K_G(X_i)$ is isomorphic to the Grothendieck ring $K_G[X]$ of the inverse spectrum X.*

PROOF. The functor K_G is a contravariant functor acting from the category Top_G of topological G-spaces to the category Ring of rings, and hence it induces a contravariant functor acting from the category $\text{Invspec}(\text{Top}_G)$ of inverse spectra of topological G-spaces to the category $\text{Dirspec}(\text{Ring})$ of direct spectra of rings. Thus, $\{K_G(X_i)\}_{i \in I}, \{K_G(p_i^{i'}): K_G(X_i) \to K_G(X_{i'})\}_{i < i' \in I}$ is a direct spectrum of rings for any inverse spectrum $(\{X_i\}_{i \in I}, \{p_i^{i'}\}_{i < i' \in I})$ of topological G-spaces. Since the completion functor K_G^n commutes with the direct limit, it follows that $\varinjlim_{i \in I} K_G(X_i) \xrightarrow{\sim} K_G[X]$. Indeed, if $(\{S_i\}_{i \in I}, \{p_i^{i'}\}_{i < i' \in I})$ is a direct spectrum of semirings, then

$$\varinjlim_{i \in I} K_G^n(S_i) = \varinjlim_{i \in I}(S_i \times S_i/\Delta(S_i)) \xrightarrow{\sim} \varinjlim_{i \in I} S_1 \times \varinjlim_{i \in I} S_i/\varinjlim_{i \in I} \Delta(S_i)$$

$$\xrightarrow{\sim} \varinjlim_{i \in I} S_i \times \varinjlim_{i \in I} S_i/\Delta\left(\varinjlim_{i \in I} S_i\right),$$

$$K_G^n\left(\varinjlim_{i \in I} S_i\right) = \varinjlim_{i \in I} S_i \times \varinjlim_{i \in I} S_i/\Delta\left(\varinjlim_{i \in I} S_i\right).$$

Accordingly, $K_G^n(\varinjlim_{i \in I} S_i) \xrightarrow{\sim} \varinjlim_{i \in I} K_G^n(S_i)$. Consequently, $\varinjlim_{i \in I} K_G(X_i) \xrightarrow{\sim} K_G[X]$.

Suppose that $(X, *) = \{(X_i, *)\}_{i \in I}$ belongs to the category of inverse spectra of topological G-spaces with distinguished point. The Grothendieck group $K_G[X, *]$ of the inverse spectrum $(X, *)$ with distinguished point is defined to be the limit of the direct spectrum of Grothendieck groups of the spaces with distinguished points: $K_G[X, *] \stackrel{\text{def}}{=} \varinjlim_{i \in I} K_G(X_i, *)$. We consider the inverse spectrum $(X, Y) \in \text{Invspec}(\text{CTop}_G^2)$ of pairs of compact G-spaces.

The Grothendieck group $K_G[X, Y]$ of an inverse spectrum of pairs of topological G-spaces is defined to be $K_G[X, Y] \stackrel{\text{def}}{=} K_G[X/Y, *]$, the Grothendieck group of the inverse spectrum $(X/Y, *)$ with distinguished element. It follows from the definition of the group $K_G[X, Y]$ that

$$K_G[X, Y] \to \varinjlim_{i \in I} K_G(X_i, Y_i).$$

We define the group $K_G^{-n}[X, Y]$ and construct an extraordinary cohomology theory on the category $\text{Invspec}(\text{CTop}_G^2)$ of inverse spectra of compact G-spaces for any $n \geq 0$.

DEFINITION 3. For any integer $n \geq 0$ and any inverse spectrum X of compact G-spaces let

$$K_G^{-n}[X] = K_G[X \times D^n, X \times S^{n-1}].$$

For an inverse spectrum (X, Y) of pairs of compact G-spaces

$$K_G^{-n}[X, Y] = K_G[X \times L^n, X \times S^{n-1} \cup Y \times D^n],$$

where S^{n-1} and D^n are, respectively, the $(n-1)$-dimensional sphere and the n-dimensional unit ball in R^n, and the group G acts on them trivially. Obviously, for any inverse spectrum $X = \{X_i\}_{i \in I}$ of G-spaces and any integer $n \geq 0$ the group $K_G^{-n}[X]$ is isomorphic to $\varinjlim_{i \in I} K_G^{-n}(X_i)$. Similarly, for any inverse spectrum $(X, Y) = \{(X_i, Y_i)\}_{i \in I}$ of pairs of compact G-spaces

$$K_G^{-n}[X, Y] \xrightarrow{\sim} \varinjlim_{i \in I} K_G^{-n}(X_i, Y_i).$$

By the periodicity theorem ([1], [5]), $\beta_i: K_G^{-(n+2)}(X_i) \xrightarrow{\sim} K_G^{-n}(X_i)$ is an isomorphism for any G-space X_i in an inverse spectrum $X = (\{X_i\}_{i \in I}, \{p_i^{i'}\}_{i < i' \in I})$ of compact G-spaces. Consequently, the system $\{\beta_i\}_{i \in I}$ of morphisms is an isomorphism of the direct spectra $\{K_G^{-(n+2)}(X_i)\}_{i \in I}$ and $\{K_G^{-n}(X_i)\}_{i \in I}$, and then $\beta = \varinjlim_{i \in I} \beta_i$ is an isomorphism of the groups $K_G^{-(n+2)}[X]$ and $K_G^{-n}[X]$. Accordingly, we have

THEOREM 3. *For an arbitrary inverse spectrum of compact G-spaces and an arbitrary integer $n \geq 0$ the groups $K_G^{-(n+2)}[X]$ and $K_G^{-n}[X]$ are isomorphic.*

It follows from the periodicity theorem that the groups $K_G^n[X]$ can be defined for all integers n, negative or positive.

THEOREM 4. *For any inverse spectrum of pairs (X, Y) of compact G-spaces there is the infinite exact sequence*

$$\begin{array}{ccc} K_G^0[X, Y] \longrightarrow K_G^0[X] \longrightarrow K_G^0[Y] \\ \uparrow & & \downarrow \\ K_G^1[Y] \longleftarrow K_G^1[X] \longleftarrow K_G^1[X, Y] \end{array}$$

PROOF. For any pair (X_i, Y_i) of spaces in the inverse spectrum $(X, Y) = (\{(X_i, Y_i)\}_{i \in I}, \{p_i^{i'}: (X_{i'}, Y_{i'}) \to (X_i, Y_i)\}_{i < i' \in I})$ we have the exact sequence

$$\cdots \to K_G^0(X_i, Y_i) \to K_G^0(X_i) \to K_G^0(Y_i)$$
$$\to K_G^1(X_i, Y_i) \to K_G^1(X_i) \to K_G^1(Y_i) \to \cdots.$$

Then the given exact sequence functor induces a contravariant functor acting from the category $\text{Invspec}(\text{CTop}_G^2)$ to the category $\text{Dirspec}(\text{ES})$ of direct spectra of exact sequences. Consequently, for $(X, Y) = (\{(X_i, Y_i)\}_{i \in I}, \{p_i^{i'}\}_{i < i' \in I})$

$$(\{\cdots \to K_G^0(X_i, Y_i) \to K_G^0(X_i) \to K_G^0(Y_i) \to K_G^1(X_i, Y_i) \to K_G^1(X_i)$$
$$\to K_G^1(Y_i) \to \cdots\}_{i \in I}, \{(p_i^{i'})^*\}_{i < i' \in I})$$

is a direct spectrum of exact sequences whose limit

$$\cdots \to \varinjlim_{i \in I} K_G^0(X_i, Y_i) \to \varinjlim_{i \in I} K_G^0(X_i) \to \varinjlim_{i \in I} K_G^0(Y_i)$$
$$\to \varinjlim_{i \in I} K_G^1(X_i, Y_i) \to \varinjlim_{i \in I} K_G^1(X_i) \to \varinjlim_{i \in I} K_G^1(Y_i) \to \cdots \quad (*)$$

is also exact [4]. If in the sequence $(*)$ we make the substitutions

$$\varinjlim_{i \in I} K_G^n(X_i, Y_i) \xrightarrow{\sim} K_G^n[X, Y], \quad \varinjlim_{i \in I} K_G^n(X_i) \xrightarrow{\sim} K_G^n[X],$$
$$\varinjlim_{i \in I} K_G^n(Y_i) \xrightarrow{\sim} K_G^n[Y],$$

then we get the exact sequence of the theorem. Let $X = (\{X_i\}_{i \in I}, \{p_i^{i'}: X_{i'} \to X_i\}_{i < i' \in I})$ and $Y = (\{Y_j\}_{j \in J}, \{p_j^{j'}: Y_{j'} \to Y_j\}_{j < j' \in J})$ be inverse spectra of compact G-spaces, and let $f = (\pi: J \to I, \{f_j: X_{\pi(j)} \to Y_j\}_{j \in J})$ and $g = (\rho: J \to I, \{g_j: X_{\rho(j)} \to Y_j\}_{j \in J})$ be G-morphisms of them.

THEOREM 5. *Homotopic G-morphisms $f, g: X \to Y$ of inverse spectra of compact G-spaces induce equal homomorphisms $K_G[f], K_G[g]: K_G[Y] \to K_G[X]$ of the Grothendieck rings.*

PROOF. Homotopic G-morphisms $f, g: X \to Y$ of inverse spectra induce canonically homotopic morphisms $f^*, g^*: \text{Vect}_G(Y) \to \text{Vect}_G(X)$ of the direct spectra of the corresponding vector G-bundles. Indeed, by the definition of the homotopy relation for the morphisms f, $g\colon X \to Y$, $\forall j \in J\ \exists i \in I$, $\pi(j) < i$, $\rho(j) < i$ such that $f_j \cdot p^i_{\pi(j)}$ is homotopic to $g_j \cdot p^i_{\rho(j)}$ [2], and then $(f_j \cdot p^i_{\pi(j)})^*(p) \approx (g_j \cdot p^i_{\rho(j)})^*(p)$ for any G-bundle $p \in \mathrm{Vect}_G(Y_j)$, i.e., $(p^i_{\pi(j)})^* \cdot f_j^* = (p^i_{\rho(j)})^* \cdot g_j^*$ in the class of isomorphic bundles ([1], [5]). But this means that the morphisms f^*, $g^*\colon \mathrm{Vect}_G(Y) \to \mathrm{Vect}_G(X)$ of the direct spectra are canonically homotopic [2]. The canonically homotopic morphisms f^* and g^* of the direct spectra induce equal morphisms $f_*^* = g_*^*$ of the limits of these direct spectra [2]. Consequently, $f_*^* = g_*^*\colon \mathrm{Vect}_G[Y] \to \mathrm{Vect}_G[X]$. But this implies that $K_G[f] = K_G[g]\colon K_G[Y] \to K_G[X]$. The theorem is proved.

We construct a special sequence for the K_G-theory on the category of inverse spectra of topological spaces. Consider an inverse spectrum $(X, F) = (\{(X_i, F_i)\}_{i\in I}, \{(p_i^{i'}, \theta_i^{i'})\colon (X_{i'}, F_{i'}) \to (X_i, F_i)\}_{i<i'\in I})$ of pairs, where X is an inverse spectrum of compact G-spaces, and $F = \{F_i\}_{i\in I}$ is an inverse spectrum of finite coverings of the spaces X_i by closed G-stable subsets: $F_i = \{F_i^{\alpha_i}\}_{\alpha_i \in S_i}$ for any $i \in I$. Since the relation $(X_i, F_i) \to W(X_i, F_i) = \{(n, x) \in N_{F_i} \times X_i;\ \nu(n)$ a face of the simplex $\xi(x)\}$, where N_{F_i} is the nerve of the covering F_i, $\nu\colon |N_{F_i}| \to N_{F_i}$, $\xi\colon X_i \to N_{F_i}$, and $\xi(x) = \{\alpha_i \in S_i;\ x \in F_i^{\alpha_i}\}$, is a covariant functor [1], it induces on the category $\mathrm{Invspec}(\mathrm{Top}_G)$ of inverse spectra of topological G-spaces a covariant functor assigning to each inverse spectrum $(X, F) = (\{(X_i, F_i)\}_{i\in I}, \{(p_i^{i'}, \theta_i^{i'})\}_{i<i'\in I})$ the inverse spectrum

$$W(X, F) = (\{W(X_i, F_i)\}_{i\in I}, \{|\theta_i^{i'}| \times p_i^{i'}\}_{i<i'\in I}).$$

The projections $\{p_{2i}\}_{i\in I}\colon \{W(X_i, F_i)\}_{i\in I} \to \{X_i\}_{i\in I}$ are morphisms of the given inverse spectra.

THEOREM 6. *The projections $\{p_{21}\}_{i\in I}\colon \{W(X_i, F_i)\}_{i\in I} \to \{X_i\}_{i\in I}$ of the inverse spectra induce a group isomorphism $K_G^*[X] \cong K_G^*[W(X, F)]$.*

PROOF. For any $i \in I$ we have that the projection $p_{2i}\colon W(X_i, F_i) \to X_i$ induces an isomorphism $K_G^*(X_i) \simeq K_G^*(W(X_i, F_i))$ [1], and then the projections $\{p_{2i}\}_{i\in I}$ induce an isomorphism of direct spectra $\{K_G^*(X_i)\}_{i\in I} \xrightarrow{\sim} \{K_G^*(W(X_i, F_i))\}_{i\in I}$. Consequently, the limits of these direct spectra are also isomorphic: $\varinjlim_{i\in I} K_G^*(X_i) \xrightarrow{\sim} \varinjlim_{i\in I} K_G^*(W(X_i, F_i))$. By what was proved above, $\varinjlim_{i\in I} K_G^*(X_i) \cong K_G^*$ and $\varinjlim_{i\in I} K_G^*(W(X_i, F_i)) \xrightarrow{\sim} K_G^*[W(X, F)]$, and hence $K_G^*[X] \cong K_G^*[W(X, F)]$.

Suppose that ${}^iE_r^{p,q}$ is a spectral sequence of type ${}^iE_2$ constructed on the component $W(X_i, F_i)$ of the inverse spectrum $W(X, F) = \{W(X_i, F_i)\}_{i\in I}$. Since a spectral sequence is functorial, it associates with the inverse spectrum $W(X, F)$ the direct spectrum $\{{}^iE_r^{p,q}\}_{i\in I}$ of bigraded modules.

DEFINITION 4. The limit $\varinjlim_{i \in I} {}^i E_r^{p,q}$ of the direct spectrum $\{{}^i E_r^{p,q}\}_{i \in I}$ of bigraded modules is called a spectral sequence of the inverse spectrum $W(X, F)$ and is denoted by $|E_r^{p,q}$. Thus, the spectral sequence defined converges to $K_G^*[W(X, F)] \xrightarrow{\sim} K_G^*[X]$; indeed, for each $i \in I$ the spectral sequence ${}^i E_r^{p,q}$ converges to $K_G^*(W(X_i, F_i))$, and then $\varinjlim_{i \in I} {}^i E_r^{p,q}$ converges to

$$\varinjlim_{i \in I} K_G^*(W(X_i, F_i)) \cong K_G^*[W(X, F)] \xrightarrow{\sim} K_G^*[X].$$

THEOREM 7. *There exists a spectral sequence $|E_r^{p,q}$ converging to $K_G^*[X]$ and having term $|E_2^{p,q}$ equal to*

$$|E_2^{p,q} = \varinjlim_{i \in I} H^p(N_{F_i}, K_G^g(F_i^r)).$$

PROOF. For any component X_i of the inverse spectrum $X = \{X_i\}_{i \in I}$ there exists a spectral sequence ${}^i E_r^{p,q}$ converging to $K_G^*(X_i)$ and having term ${}^i E_2^{p,q}$ equal to

$$^i E_2^{p,q} = H^p(N_{F_i}, K_G^q(F_i^\sigma)).$$

This implies that $\varinjlim_{i \in I} {}^i E_r^{p,q} = |E_r^{p,q}$ converges to $\varinjlim_{i \in I} K_G^*(X_i) \xrightarrow{\sim} K_G^*[X]$, and

$$\|E_2^{p,q} = \varinjlim {}^i E_2^{p,q} = \varinjlim_{i \in I} H^p(N_{F_i}, K_G^q(F_i^\sigma)).$$

We consider the category with objects the pairs (X, G), where $X = (\{X_i\}_{i \in I}, \{p_i^{i'}\}_{i < i' \in I})$ is an inverse spectrum of G-spaces, and with morphisms the triples $(\pi, \theta, f): (Y, H) \to (X, G)$, $Y = (\{Y_j\}_{j \in J}, \{p_j^{j'}\}_{j > j' \in J})$, where $\pi: I \to J$ is an isotone mapping of directed sets, θ is a group homomorphism $H \to G$, $f = (\pi: I \to J, \{f_i: Y_{n(i)} \to X_i\}_{i \in I})$, and θ is an equivariant mapping, that is, $f_i(h \cdot y) = \theta(h) f_i(y)$ for any $i \in I$, $h \in H$, $y \in Y_{n(i)}$.

Let $\text{Vect}_G(X)$ be the category of vector G-bundles over a G-space X. The morphism $(\theta, f): (Y, H) - (X, G)$ induces a morphism $f^*: \text{Vect}_G(X) \to \text{Vect}_H(Y)$ of categories of vector bundles that assigns to each G-bundle over X the induced H-bundle over Y. Thus, the functor $K_G(X)$ determines a contravariant functor from the category of topological spaces with continuous actions of groups into the category of commutative rings. The latter functor, in turn, induces a contravariant functor acting from the aforementioned category of inverse spectra of topological spaces into the category of commutative rings with identity.

The author expresses deep gratitude to his scientific advisor M. R. Bunyatov for the statement of the problem and for help with this work.

Bibliography

1. M. Atiyah, *K-theory*, Lecture Notes, Harvard Univ., Cambridge, MA, 1964.
2. M. R. Bunyatov and S. A. Baĭramov, *K-theory on the category of Boolean algebras with closure*, Akad. Nauk Azerbaĭdzhan. SSR Dokl. **33** (1977), no. 12, 3–7; English transl. in this volume.
3. Samuel Eilenberg and Norman Steenrod, *Foundations of algebraic topology*, Princeton Univ. Press, Princeton, NJ, 1952.
4. Dale Husemoller, *Fiber bundles*, McGraw-Hill, New York, 1966.

Azerbaijan State University

Received 23/DEC/77

Translated by H. H. McFADEN

K-Theory on the Category of Distributive Lattices

UDC 513.831.513.836

M. R. BUNYATOV AND S. A. BAĬRAMOV

> Abstract. The K-functor is introduced on the category of distributive lattices with a pair of filters using the nerve by P. S. Alexandrov, which satisfies all the axioms of Steenrod and Eilenberg, except the dimension axiom.

Atiyah and Hirzebruch employed the Grothendieck group $K(X)$ to construct a certain extraordinary cohomology theory on the category of finite polyhedra, using the suspension functor. Atiyah later [1] extended the topological K-functor to the category of locally finite polyhedra. In [2] Bukhshtaber and Mishchenko constructed a K-theory on the category of infinite complexes.

In connection with these investigations the problem naturally arises of constructing a K-theory as an extraordinary cohomology theory on the maximally admissible category for cohomology theory—on the category of all topological spaces.

With this aim a K-theory is constructed on the category of distributive lattices and their homomorphisms, which gives a certain extraordinary cohomology theory on the subcategory formed by the topological spaces.

Denote by Z the category of all distributive lattices with 0 and 1 and their homomorphisms.

For any lattice[1] L denote by $\mathrm{Cov}_f(L)$, $\mathrm{Cov}_{lf}(L)$, and $\mathrm{Cov}(L)$ the sets formed by the finite, locally finite, and arbitrary coverings, respectively, of the unit of L, directed in the Moore-Smith sense with respect to the relation of refinement of coverings.

1991 *Mathematics Subject Classification.* Primary 55N15; Secondary 18F25.
Translation of Akad. Nauk Azerbaidzhan. SSR Dokl. **39** (1983), no. 5, 7–11.

[1] Unless otherwise stated, a lattice is assumed below to be a distributive lattice with 0 and 1.

For any $\alpha \in \mathrm{Cov}(L)$, $\alpha = \{\alpha(i)\}_{i \in I}$, denote by $\mathrm{nerv}\,\alpha$ the abstract simplicial complex $\{\sigma \subset I | \bigwedge_{i \in \sigma} \alpha(i) \neq 0\}$ with set I of vertices, and by $|\mathrm{nerv}\,\alpha|$ the body of this complex.

Let K_A be the Atiyah functor on the category P_f of finite polyhedra. For each finite covering $\alpha \in \mathrm{Cov}_f(L)$ let $K_A(\alpha) \stackrel{\mathrm{def}}{=} K_A(|\mathrm{nerv}\,\alpha|)$. For any coverings $\alpha, \beta \in \mathrm{Cov}_f(L)$ with $\beta > \alpha$ (β a refinement of α) denote by $K_\alpha^\beta : K_A(\alpha) \to K_A(\beta)$ the ring homomorphism of $K_A(\alpha)$ into $K_A(\beta)$ induced by some projection (refinement mapping) of the covering β into the covering α. Because of the homotopy invariance of the ring $K_A(X)$, the homomorphism $K_\alpha^\beta : K_A(\alpha) \to K_A(\beta)$ is well defined.

LEMMA 1. *The pair* $\mathrm{spec}\,L = (\{K_A(\alpha)\}_{\alpha \in \mathrm{Cov}_f(L)}, \{K_\alpha^\beta\}_{\alpha < \beta \in \mathrm{Cov}_f(L)})$ *is a direct spectrum of rings.*

DEFINITION 1. The ring $K_A(L) = \lim_{\alpha \in \mathrm{Cov}_f(L)} K_A(\alpha)$ is called the Atiyah-Grothendieck ring of the lattice L.

We now define the induced homomorphism. Let $f: L \to L'$ be a lattice homomorphism. The induced isotone mapping $f_* : \mathrm{Cov}(L) \to \mathrm{Cov}(L')$ is defined by the formula $f_*(\alpha) = \{f(\alpha(i))\}$, $i \in \mathrm{supp}\,f_*(\alpha)$, where $\mathrm{supp}\,f_*(\alpha) = \{i \in I | f(\alpha(i)) \neq 0\}$. It is easy to see that $f_*(\mathrm{Cov}_f(L)) \subset \mathrm{Cov}_f(L')$ and $f_*(\mathrm{Cov}_{lf}(L)) \subset \mathrm{Cov}_{lf}(L')$.

For each $\alpha \in \mathrm{Cov}(L)$ let $i_{\alpha,f} : |\mathrm{nerv}\,f_*(\alpha)| \to |\mathrm{nerv}\,\alpha|$ be the canonical imbedding of the subcomplex $|\mathrm{nerv}\,f_*(\alpha)|$ in the complex $|\mathrm{nerv}\,\alpha|$. Then the pair $\mathrm{spec}\,f = (f_* : \mathrm{Cov}_f(L) \to \mathrm{Cov}_f(L'), \{i_{\alpha,f}^*\}_{\alpha \in \mathrm{Cov}_f(L)})$ is a morphism of the direct spectrum $\mathrm{spec}\,L$ into the direct spectrum $\mathrm{spec}\,L'$.

The ring homomorphism of $K_A(L)$ into $K_A(L')$ induced by the morphism $\mathrm{spec}\,f : \mathrm{spec}\,L \to \mathrm{spec}\,L'$ is called the induced homomorphism. Denote this homomorphism by $f_{**} : K_A(L) \to K_A(L')$.

THEOREM 1. *The relation* $L \to K_A(L)$, $f \to f_{**} : K_A(L) \to K_A(L')$ *is a covariant functor from the category* Z *of distributive lattices with* 0 *and* 1 *into the category of associative commutative rings with identity.*

In the category Z we introduce a homotopy relation for lattice homomorphisms.

DEFINITION 2. Lattice homomorphisms $f, g : L \to L'$ are said to be homotopic if $\forall \alpha \in \mathrm{Cov}_f(L)\,\exists \beta \in \mathrm{Cov}_f(L')$ such that $\beta > f_*(\alpha)$ and $\beta > g_*(\alpha)$, and for any projections p and q of the covering β into the respective coverings $f_*(\alpha)$ and $g_*(\alpha)$ the mappings $i_{\alpha,f} \cdot |p|, i_{\alpha,g} \cdot |q| : |\mathrm{nerv}\,\beta| \to |\mathrm{nerv}\,\alpha|$, are homotopic. If homomorphisms f and g are homotopic, then we write $f \sim g$.

THEOREM 2. *The homotopy relation between lattice homomorphisms is an equivalence relation. It is natural for composition of homomorphisms.*

THEOREM 3. *The K_A-functor on the category Z is homotopy invariant: if $f, g: L \to L'$ are homotopic homomorphisms, then the induced homomorphisms $f_{**}, g_{**}: K_A(L) \to K_A(L')$ coincide, i.e., $f \sim g \to f_{**} = g_{**}$.*

To define $K_A^{-n}(L)$ we first set $K_A^{-n}((\alpha) = K_A^{-n})(|\text{nerv } \alpha|)$ for any $\alpha \in \text{Cov}_f(L)$. By definition,

$$K_A^{-n}(L) \stackrel{\text{def}}{=} \lim_{\alpha \in \text{Cov}_f(L)} K_A^{-n}(\alpha).$$

Thus, the covariant functor $K_A^{-n}: Z \to \text{Ring}$ is defined for each natural number $n \geq 0$.

We consider the pairs (L, F), where L is a lattice and F either a filter in L or L itself. A covering of a pair (L, F) is defined to be a pair (α, I') formed from a covering $\alpha \in \text{Cov}(L)$, $\alpha = \{\alpha(i)\}_{i \in I}$, and the restriction of it to a subset $I' \subset I$ such that $\bigvee_{i \in I'} \alpha(i) \in F$. Let $(\beta, J') \in \text{Cov}(L, F)$, where $\beta = \{\beta(i)\}_{i \in J}$, $J' \subset J$. A mapping $p: (J, J') \to (I, I')$ is called a projection of the covering (β, J') into the covering (α, I') if it is a projection of the covering β into the covering α. A covering (β, J') is a refinement of a covering (α, J') if β is a refinement of α. The relation of (β, J') being a refinement of (α, I') is denoted by $(\beta, J') > (\alpha, I')$.

Thus, the class $\text{Cov}(L, F)$ of all coverings of a pair (L, F) forms a directed set in the Moore-Smith sense. The respective classes $\text{Cov}_f(L, F)$ and $\text{Cov}_{lf}(L, F)$ of finite and locally finite coverings are defined in the natural way.

For a covering $(\alpha, I') \in \text{Cov}(L, F)$ we define $\text{nerv}(\alpha, I')$ as the pair $(\text{nerv } \alpha, \text{nerv } \alpha | I')$ of complexes formed from the nerve of the covering α and from the subcomplex of it that is the nerve of the restriction $\alpha | I'$ of this covering α. For each natural number $n \geq 0$ and for each covering $(\alpha, I') \in \text{Cov}_f(L, F)$ we let $K_A^{-n}(\alpha, I') \stackrel{\text{def}}{=} K_A^{-n}(|\text{nerv } \alpha|, |\text{nerv } \alpha | I'|)$. For any coverings $(\alpha, I'), (\beta, J') \in \text{Cov}_f(L, F)$ such that $(\beta, J') > (\alpha, I')$ denote by $K_{(\alpha, I')}^{(\beta, J')}: K_A^{-n}(\alpha, I') \to K_A^{-n}(\beta, J')$ the ring homomorphism induced by some projection. It is easy to see that

$$\text{spec}^{-n}(L, F) = (\{K_A^{-n}(\alpha, I')\}_{(\alpha, I') \in \text{Cov}_f(L, F)}, \{K_{(\alpha, I')}^{(\beta, J')}\}_{(\alpha, I') < (\beta, J')})$$

is a direct spectrum of rings.

DEFINITION 3. *The limit*

$$K_A^{-n}(L, F) = \lim_{(\alpha, I') \in \text{Cov}_f(L, F)} K_A^{-n}(\alpha, I')$$

of the direct spectrum $\text{spec}^{-n}(L, F)$ is called the n-dimensional Atiyah-Grothendieck ring of the lattice with filter (L, F).

A homomorphism of a pair (L, F) to a pair (L', F'), where F is a filter in the lattice L and F' is a filter in the lattice L', is defined to be a lattice homomorphism $f: L \to L'$ such that $f(F) \subset F'$.

For each covering $(\alpha, I') \in \text{Cov}(L, F)$ and each homomorphism $f: (L, F) \to (L', F')$ of pairs we define a mapping $f_*: \text{Cov}(L, F) \to \text{Cov}(L', F')$ by the formula $f_*(\alpha, I') = (f_*(\alpha), f^*(I'))$, where $f_*(I') = I' \cap \text{supp}\, f_*(\alpha)$. The mapping f_* is an isotopic mapping.

Let $i_{f,(\alpha, I')}: (|\text{nerv}\, f_*(\alpha)|, |\text{nerv}\, f_*(\alpha)|f_*(I')|) \to |\text{nerv}\,\alpha|, |\text{nerv}\,\alpha|I'|)$ be an imbedding map of pairs of complexes.

The pair $(f_*: \text{Cov}_f(L, F) \to \text{Cov}_f(L', F'), \{i^*_{f,(\alpha, I')}\}_{(\alpha, I') \in \text{Cov}_f(L, F)})$ is a morphism of the direct spectrum $\text{spec}^{-n}(L, F)$ to the direct spectrum $\text{spec}^{-n}(L', F')$. This morphism induces a homomorphism of limit rings. Denote this homomorphism by $f_{**}: K_A^{-n}(L, F) \to K_A^{-n}(L, F')$.

THEOREM 4. *Let ZF be the category of pairs (L, F) formed by a lattice and a filter, and of all possible homomorphisms of them. Then for each natural number $n \geq 0$ the correspondence $(L, F) \to K_A^{-n}(L, F)$, $f \to f_{**}: K_A^{-n}(L, F) \to K_A^{-n}(L', F')$ is a covariant functor from the category ZF to the category of rings.*

In ZF, just as in the category Z, we introduce the homotopy relation with respect to which the functor $K_A^{-n}(L, F)$ is a homotopy invariant.

To construct an exact cohomology sequence it is first necessary to modify the category ZF by imbedding it in the category ZF^2. The objects in ZF^2 are ordered triples (L, Φ, ψ), where L is a lattice, Φ a filter in the lattice, and ψ either L itself or a finer filter than Φ.

A morphism of the triple (L, Φ, ψ) to the triple (L', Φ', ψ') is defined to be a lattice homomorphism $f: L \to L'$ that is a mapping $f: (L, \Phi, \psi) \to (L', \Phi', \psi')$ of triples.

DEFINITION 4. Let Φ be a filter. A family $\{\alpha(i)\}_{i \in I}$ of elements of L is called a covering of Φ if the least upper bound of this family belongs to Φ: $\bigvee_{i \in I} \alpha(i) \in \Phi$.

It is easy to see that the class $\text{Cov}(\Phi)$ of all coverings of the filter Φ forms a directed set in the Moore-Smith sense with respect to refinement. More generally, a family $\{\alpha(i)\}_{i \in I}$ of elements together with a distinguished subset of indices $I' \subset I$ is called a covering of the pair (Φ, ψ) if $\alpha \in \text{Cov}(\Phi)$ and $\alpha|I' \in \text{Cov}(\psi)$. The class of all coverings of the pair (Φ, ψ) is denoted by $\text{Cov}(\Phi, \psi)$; $\text{Cov}(\Phi, \psi)$ is a directed set with respect to refinement when Φ is a filter and ψ is either a filter or the lattice L itself.

The ring $K_A^{-n}(L, \Phi, \psi) \stackrel{\text{def}}{=} \lim_{(\alpha, I') \in \text{Cov}_f(\Phi, \psi)} K_A^{-n}(|\text{nerv}(\alpha, I')|)$ is defined by analogous arguments for an ordered triple $(L, \Phi, \psi) \in ZF^2$.

The induced homomorphisms of the rings $K_A^{-n}(L, \Phi, \psi)$ are defined similarly.

Thus, for each natural number $n \geq 0$ the covariant functor $K_A^{-n}(L, \Phi, \psi)$ is defined from the category ZF^2 to the category of rings. Further, two morphisms $f, g: (L, \Phi, \psi) \to (L', \Phi', \psi')$ are said to be homotopic if they are homotopic as lattice homomorphisms.

To construct an exact cohomology sequence we use the exact cohomology sequences $K^*(\alpha, I')$ of the pairs $(|\operatorname{nerv}\alpha|, |\operatorname{nerv}\alpha|I'|)$, where $(\alpha, I') \in \operatorname{Cov}_f(\Phi, \psi)$.

We define the projections $\operatorname{pr}_1 : \operatorname{Cov}(\Phi, \psi) \to \operatorname{Cov}(\Phi)$ and $\operatorname{pr}_2 : \operatorname{Cov}(\Phi, \psi) \to \operatorname{Cov}(\psi)$ by the formulas $\operatorname{pr}_1(\alpha, I') = \alpha$ and $\operatorname{pr}_2(\alpha, I') = \alpha|I'$. It is clear that these mappings are isotone epimorphisms. Denote by $K^{-n}_{(\Phi, \psi)}(L, \Phi, L)$ the limit of the direct spectrum

$$(\{K^{-n}_A(\operatorname{pr}_1(\alpha, I'))\}|_{(\alpha, I') \in \operatorname{Cov}_f(\Phi, \psi)}, \{\kappa^{\operatorname{pr}_1(\beta, J')}_{\operatorname{pr}_1(\alpha, I')}\}_{(\alpha, I')<(\beta, J')})$$

induced by the direct spectrum $(\{K^{-n}_A(\alpha)\}_{\alpha \in \operatorname{Cov}_f(\Phi)}, \{\kappa^\beta_\alpha\}_{\alpha<\beta})$ with respect to the isotone mapping pr_1. Thus,

$$K^{-n}_{(\Phi, \psi)}(L, \Phi, L) \stackrel{\text{def}}{=} \lim_{(\alpha, I') \in \operatorname{Cov}_f(\Phi, \psi)} K^{-n}_A(\operatorname{pr}_1(\alpha, I')).$$

An analogous definition applies to the ring

$$K^{-n}_{(\Phi, \psi)}(L, \psi, L) \stackrel{\text{def}}{=} \lim_{(\alpha, I') \in \operatorname{Cov}_f(\Phi, \psi)} K^{-n}_A(\operatorname{pr}_2(\alpha, I')).$$

The projections pr_1 and pr_2 are epimorphisms, so we have the canonical isomorphisms

$$\begin{aligned} \operatorname{pr}_{1**} : K^{-n}_{(\Phi, \psi)}(L, \Phi, L) \stackrel{\sim}{\to} K^{-n}_A(L, \Phi, L), \\ \operatorname{pr}_{2**} : K^{-n}_{(\Phi, \psi)}(L, \psi, L) \stackrel{\sim}{\to} K^{-n}_A(L, \psi, L). \end{aligned} \quad (1)$$

Since the limit of a direct spectrum of exact sequences is exact, we get an exact sequence

$$\cdots \to K^{-1}_A(L, \Phi, \psi) \stackrel{j^*(\Phi, \psi)}{\longrightarrow} K^{-1}_{(\Phi, \psi)}(L, \Phi, L) \stackrel{i^*(\Phi, \psi)}{\longrightarrow} K^{-1}_{(\Phi, \psi)}(L, \psi, L)$$
$$\stackrel{\delta(\Phi, \psi)}{\longrightarrow} K^0_A(L, \Phi, \psi) \stackrel{j^*(\Phi, \psi)}{\longrightarrow} K^0_{(\Phi, \psi)}(L, \Phi, L) \stackrel{i^*(\Phi, \psi)}{\longrightarrow} K^0_{(\Phi, \psi)}(L, \psi, L) \quad (2)$$

by passing to the limit of the direct spectrum $\{K^*(\alpha, I')\}_{(\alpha, I') \in \operatorname{Cov}_f(\Phi, \psi)}$.

Taking the isomorphisms (1) into account, we get from this sequence the exact sequence

$$\begin{aligned} \cdots \to K^{-1}_A(L, \Phi, \psi) \stackrel{j^*}{\to} K^{-1}_A(L, \Phi, L) \stackrel{i^*}{\to} K^{-1}_A(L, \psi, L) \\ \stackrel{\delta}{\to} K^0_A(L, \Phi, \psi) \stackrel{j^*}{\to} K^0_A(L, \Phi, L) \stackrel{i^*}{\to} K^0_A(L, \psi, L), \end{aligned} \quad (3)$$

where $j^* = \operatorname{pr}_{1**} \cdot j^*_{(\Phi, \psi)}$, $i^* = \operatorname{pr}_{2**} \cdot i^*_{(\Phi, \psi)} \cdot \operatorname{pr}^{-1}_{1**}$, and $\delta = \delta_{(\Phi, \psi)} \cdot \operatorname{pr}^{-1}_{2**}$.

THEOREM 6 (Exactness Axiom). *For each triple (L, Φ, ψ) there exists an exact cohomology sequence of the form* (3).

To construct an exact cohomology sequence for the objects in the category ZF, the latter is imbedded in the category ZF^2 by means of the correspondence $(L, F) \to (L, \{1_2\}, F)$.

Bibliography

1. M. Atiyah and F. Hirzebruch, *Vector bundles and homogeneous spaces*, Proc. Sympos. Pure Math., vol. 3, Amer. Math. Soc., Providence, RI, 1961, pp. 7–38.
2. V. M. Bukhshtaber and A. S. Mishchenko, *A K-theory on the category of infinite cell complexes*, Izv. Akad. Nauk SSSR Ser. Mat. **32** (1968), no. 3, 560–604; English transl. in Math. USSR-Izv. **2** (1968).
3. Samuel Eilenberg and Norman Steenrod, *Foundations of algebraic topology*, Princeton Univ. Press, Princeton, NJ, 1952.

Azerbaijan State University

Received 16/NOV/81

Translated by H. H. McFADEN

K-Theory on the Category of Topological Spaces

UDC 513.836

M. R. BUNYATOV AND S. A. BAĬRAMOV

Abstract. Using the nerve by P. S. Alexandrov on the category of topological spaces, an extraordinary theory of cohomologies is constructed. On the category of finite polyhedron this theory coincides with the K-theory by Atiyah.

In [2] we constructed a K-theory on the category of distributive lattices. In the present article we use the indicated K-functor to construct a K-theory on the category of all topological spaces that coincides on the category of finite polyhedra with the K-theory of Atiyah [1].

We remark that while the Atiyah K-functor is connected with the singular cohomology functor, the K-functor introduced by us is connected with the spectral cohomology functor.

Suppose that X is a topological space, Open X is its lattice of open subsets, $A \subset X$, and $B(A)$ is the filter in Open X formed by the open neighborhoods of A.

We "imbed" the category of pairs (X, A) of topological spaces in the category ZF^2 of lattices with a pair of filters, via the correspondence $(X, A) \to$ (Open X, $\{X\}$, $B(A)$). It is clear that this correspondence is a contravariant functor from the category Top^2 to the category ZF^2; here, to each continuous mapping $f: (X, A) \to (Y, B)$ of pairs there corresponds the homomorphism

$$f^*: (\text{Open } Y, \{Y\}, B(B)) \to (\text{Open } X, \{X\}, B(A))$$

of triples of the category ZF^2 induced by it.

For each integer $n \geq 0$ we define a functor $K_{\text{spec}}^{-n}: \text{Top}^2 \to \text{Ring}$ on the

---

1991 *Mathematics Subject Classification.* Primary 55N15, 55N20.
Translation of Akad. Nauk Azerbaidzhan. SSR Dokl. **39** (1983), no. 6, 14–18.

category Top by the formula $\forall (X, A) \in \text{Top}^2$

$$K_{\text{spec}}^{-n}(X, A) \overset{\text{def}}{=} K^{-n}(\text{Open } X, \{X\}, B(A)).$$

Denote by f^{**} the homomorphism of the ring $K_{\text{spec}}^{-n}(Y, B)$ into the ring $K_{\text{spec}}^{-n}(X, A)$ that is induced by a continuous mapping $f: (X, A) \to (Y, B)$ of pairs of topological spaces.

The following theorem is a consequence of the covariance of the functor $K^{-n}: LF^2 \to \text{Ring}$ from the category LF^2 and the contravariance of the correspondence $(X, A) \to (\text{Open } X, \{X\}, B(A))$.

THEOREM 1. *The correspondence* $(X, A) \to K_{\text{spec}}^{-n}(X, A)$, $f \to f^{**}$ *is a contravariant functor from the category* Top^2 *of pairs of topological spaces into the category of rings.*

In particular, for a space X the ring $K_{\text{spec}}^{-n}(X)$ is defined as follows: $K_{\text{spec}}^{-n}(X) \overset{\text{def}}{=} K^{-n}(\text{Open } X, \{X\}, \text{Open } X)$. Then for a space $A \subset X$ we get that

$$K_{\text{spec}}^{-n}(A) = K^{-n}(\text{Open } A, \{A\}, \text{Open } A),$$

where $\text{Open } A$ is the lattice of open subsets of A.

We have the easily established canonical isomorphism

$$\chi_{(X,A)}: K^{-n}(\text{Open } X, B(A), \text{Open } X) \to K_{\text{spec}}^{-n}(A).$$

This canonical isomorphism is constructed from the canonical isomorphism $\text{Cov}_f(B(A))$ onto $\text{Cov}_f(A)$.

We endow the category Top^2 of pairs of topological spaces with the homotopy relation defined as follows:

DEFINITION 1. Continuous mappings $f, g: (X, A) \to (Y, B)$ of pairs of topological spaces are said to be nerve-homotopic if the homomorphisms $f^*, g^*: (\text{Open } Y, \{Y\}, B(B)) \to (\text{Open } X, \{X\}, B(A))$ of triples of the category LF^2 induced by them are homotopic homomorphisms. The relation of mappings f and g being nerve homotopic is denoted by f, n, g.

Obviously, this homotopy relation is an equivalence relation for continuous mappings of pairs of topological spaces.

In view of the homotopy invariance of the functor K^{-n} on the category LF^2 we have

THEOREM 2. *Nerve-homotopic mappings* $f, g: (X, A) \to (Y, B)$ *of pairs of topological spaces induce equal homomorphisms* $f^{**}, g^{**}: K_{\text{spec}}^{-n}(Y, B) \to K_{\text{spec}}^{-n}(X, A)$ *of rings, i.e.,* $f^n g \to f^{**} = g^{**}$.

We now construct an exact cohomology sequence for a pair (X, A) of topological spaces.

For the triple (Open X, $\{X\}$, $B(A)$) corresponding to a pair (X, A) we have the following commutative diagram:

$$\xrightarrow{i^*_{(X,A)}} K^{-1}(\text{Open } A, B(A), \text{Open } A) \xrightarrow{\delta(X,A)}$$

$$\uparrow \chi(X,A) \qquad\qquad\qquad\qquad\qquad\qquad (*)$$

$$\cdots \to K^{-1}(\text{Open } X, \{X\}, \text{Open } X) \xrightarrow{i^*} K^{-1}(\text{Open } X, B(A), \text{Open } X)$$

$$\xrightarrow{\delta} K^0(\text{Open } X, \{X\}, B(A)) \to \cdots ,$$

where the lower row of this diagram is the exact cohomology sequence of the triple (Open X, $\{X\}$, $B(A)$). Making the substitutions

$$i^*_{(X,A)} = \chi_{(X,A)} \cdot i^*, \qquad \delta_{(X,A)} = \delta \cdot \chi^{-1}_{(X,A)}$$

in the diagram $(*)$, we get the exact sequence

$$\cdots \to K^{-1}(\text{Open } X, \{X\}, \text{Open } X) \xrightarrow{i^*_{(X,A)}} K^{-1}(\text{Open } A, B(A), \text{Open } A)$$

$$\xrightarrow{\delta_{(X,A)}} K^0(\text{Open } X, \{X\}, B(A)) \to \cdots .$$

Using the definitions of the rings $K^{-n}_{\text{spec}}(X, A)$, $K^{-n}_{\text{spec}}(X)$, and $K^{-n}_{\text{spec}}(A)$, we now have the exact sequence

$$\cdots \to K^{-1}_{\text{spec}}(X, A) \xrightarrow{j^*_{(X,A)}} K^{-1}_{\text{spec}}(X) \xrightarrow{i^*_{(X,A)}} K^{-1}_{\text{spec}}(A) \xrightarrow{\delta_{(X,A)}} K^0_{\text{spec}}(X, A)$$

$$\xrightarrow{j^*_{(X,A)}} K^0_{\text{spec}}(X) \xrightarrow{i^*_{(X,A)}} K^0_{\text{spec}}(A). \qquad (**)$$

This sequence is called the exact cohomology sequence of the pair (X, A) of topological spaces.

It is easy to see that the homomorphism $i^*_{(X,A)} : K^{-n}_{\text{spec}}(X) \to K^{-n}_{\text{spec}}(A)$ coincides with the homomorphism induced by the imbedding $\text{inn}_{A \in X} : A \to X$ of the subspace A in the space X, i.e.,

$$i^*_{(X,A)} = \text{inn}^*_{A \in X}.$$

Thus, the sequence $(**)$ is in fact a cohomology sequence for the pair (X, A).

Accordingly, we have proved the following theorem.

THEOREM 3. *An arbitrary pair (X, A) of topological spaces has the exact cohomology sequence $(**)$.*

Thus, summing up everything, we have

THEOREM 4. *The system $\{\{K^{-n}_{\text{spec}}(X, A)\}_{n \geq 0}, \delta_{(X,A)}\}_{(X,A) \in \text{Top}^2}$ of functors and natural transformations of them satisfies all the axioms of a cohomology theory on the maximal category Top^2 of pairs of topological spaces, with the exception of the dimension axiom.*

We show that the spectral K-functor on the category of finite polyhedra coincides with the Atiyah-Hirzebruch K-functor. With this goal we use the

Spanier canonical mappings [3]. A continuous mapping $f: X \to |\operatorname{nerv} \alpha|$ satisfying the condition $f^{-1}(\operatorname{st} U) \subset U$ is called a canonical mapping. It is known that for a locally finite open covering α any two canonical mappings $f, g: X \to |\operatorname{nerv} \alpha|$ are homotopic. In the case of paracompact spaces, and hence finite polyhedra, for any open covering α the set of canonical mappings $f: X \to |\operatorname{nerv} \alpha|$ is nonempty. For any projection $p: \beta \to \alpha$ of open coverings and any canonical mapping $f: X \to |\operatorname{nerv} \beta|$ the composition $f \cdot |p|: X \to |\operatorname{nerv} \alpha|$ is a canonical mapping.

Let K be a finite symplicial complex, and let $\operatorname{st} K$ be the covering of the stars $\operatorname{st}_K P$ of all the vertices $P \in K^0$ of the complex K.

Denote by $\operatorname{st}: |K| \to |\operatorname{nerv} \operatorname{st}|$ the canonical homeomorphism $\operatorname{st}(P) = \operatorname{st}_K(P) \, \forall P \in K^0$. The mapping st is a canonical mapping.

Indeed, the inverse image $\operatorname{st}^{-1}(\operatorname{st}_{\operatorname{nerv} \operatorname{st} K} \operatorname{st}_K P)$ of the open star $\operatorname{st}_{\operatorname{nerv} \operatorname{st} K} \operatorname{st}_K P$ with respect to the complex $\operatorname{nerv} \operatorname{st} K$ of the star $\operatorname{st}_K P \in \operatorname{nerv}^0 \operatorname{st} K$ of an arbitrary vertex $P \in K^0$ coincides with the star $\operatorname{st}_K P$ itself, i.e.,

$$\operatorname{st}^{-1}(\operatorname{st}_{\operatorname{nerv} \operatorname{st} K} \operatorname{st}_K P) = \operatorname{st}_K P.$$

For each finite open covering $\alpha \in \operatorname{Cov}_f(K)$ we denote by $\alpha_*: K^{-n}(|\operatorname{nerv} \alpha|) \to K^{-n}(|K|)$ the homomorphism induced by some canonical mapping $f: |K| \to |\operatorname{nerv} \alpha|$. The family $\{\alpha_*\}_{\alpha \in \operatorname{Cov}_f(X)}$ is a morphism of the direct spectrum

$$(\{K^{-n}(|\operatorname{nerv} \alpha|)\}_{\alpha \in \operatorname{Cov}_f(|K|)}, \{K_\alpha^\beta\}_{\alpha < \beta \in \operatorname{Cov}_f(|K|)}) \quad \text{in the ring } K^{-n}(|K|).$$

Indeed, for any coverings $\alpha, \beta \in \operatorname{Cov}_f(|K|)$ such that $\beta > \alpha$ the following diagram is commutative:

$$\begin{array}{ccc} K^{-n}(|\operatorname{nerv} \alpha|) & \xrightarrow{\alpha_X} & \\ \downarrow{\scriptstyle K_\alpha^\beta} & & K^{-n}(|K|) \\ K^{-n}(|\operatorname{nerv} \beta|) & \xrightarrow{\beta_X} & \end{array}$$

The homomorphism induced by this morphism from the ring $K_{\operatorname{spec}}^{-n}(|K|)$ into the ring $K^{-n}(|K|)$ is denoted by $\operatorname{nerv}_{|K|}^*: K_{\operatorname{spec}}^{-n}(|K|) \to K^{-n}(|K|)$. We also consider the homomorphism $(\operatorname{st}^{-1})_{|K|}^*: K_s^{-n}(|K|) \to K_{\operatorname{spec}}^{-n}(|K|)$ induced by the homomorphism $\operatorname{st}^{-1}: |\operatorname{nerv} \operatorname{st}| \to |K|$.

The family $\operatorname{nerv}^* = \{\operatorname{nerv}_x^*\}_{x \in P_f}$, $(\operatorname{st}^{-1})^* = \{(\operatorname{st}^{-1})_x^*\}_{X \in P_f}$ represents natural transformations of the functors $K_{\operatorname{spec}}^{-n}(X)$ and $K^{-n}(X)$ into each other on the category P_f of finite polyhedra.

For any finite polyhedron X the homomorphisms nerv_x^* and $(\operatorname{st}^{-1})_x^*$ are mutually inverse, and hence the natural transformations nerv^* and $(\operatorname{st}^{-1})^*$ are natural equivalences.

Thus, the theorem is proved.

THEOREM 5. *The functors K^{-n} and K_{spec}^{-n} on the category P_f of finite polyhedra are isomorphic.*

If X and Y are compact Hausdorff spaces, then the set $\text{Cov}(X) \times \text{Cov}(Y)$ is a cofinal subset of the directed set $\text{Cov}(X \times Y)$ [4], and hence to define the ring $K_{\text{spec}}(X \times Y)$ it suffices to consider the directed set $\text{Cov}_f(X) \times \text{Cov}_f(Y)$. Thus,

$$K_{\text{spec}}(X \times Y) = \lim_{\alpha = (\alpha_1, \alpha_2) \in \text{Cov}_f(X \times Y)} K(|\text{nerv}\,\alpha|)$$

$$= \lim_{(\alpha_1, \alpha_2) \in \text{Cov}_f(X \times Y)} K(|\text{nerv}\,\alpha_1 \triangle \text{nerv}\,\alpha_2|)$$

$$\approx \lim_{(\alpha_1, \alpha_2) \in \text{Cov}_f(X \times Y)} K_A(|\text{nerv}\,\alpha_1| \times |\text{nerv}\,\alpha_2|).$$

If for each $\alpha = (\alpha_1, \alpha_2) \in \text{Cov}_f(X \times Y)$

$$[p_{1\alpha}, p_{2\alpha}]: K_A(|\text{nerv}\,\alpha_1|) \times K_A(|\text{nerv}\,\alpha_2|) \to K_A(|\text{nerv}\,\alpha_1| \times |\text{nerv}\,\alpha_2|)$$

is exterior K-multiplication, then in view of its naturalness the system $\{[p_{1\alpha}, p_{2\alpha}]\}_{\alpha \in \text{Cov}_f(X \times Y)}$ of homomorphisms is a morphism of direct spectra $\{K_A(\alpha_1) \times K_A(\alpha_2)\}_{(\alpha_1, \alpha_2) \in \text{Cov}_f X \times Y} \to \{K_1(\alpha)\}_{\alpha \in \text{Cov}_f(X \times Y)}$ and thus induces a homomorphism of limit groups

$$[p_1, p_2]: K_{\text{spec}}(X) \times K_{\text{spec}}(Y) \to K_{\text{spec}}(X \times Y),$$

which we call exterior K-multiplication.

For the spectral K-functor on the category of compact Hausdorff spaces we now prove the Bott periodicity theorem, i.e., we prove that the exterior K-multiplication is an isomorphism when $Y = S^2$.

Since S^2 can be triangulated, and the triangulation $T = \{t, K\}$ is finite, the coverings $\{{}^n\tau\}$ form a cofinal subset of the set $\text{Cov}(S^2)$; here ${}^n\tau$ is the covering of S^2 associated with the n-fold barycentric subdivision ${}^nT = \{{}^nt, {}^nK\}$ of the triangulation $T = \{t, K\}$ [4]. For the covering $\tau = \{\tau_A = (\text{st}(A))\}_{A \in K}$ associated with the triangulation $T = \{t, K\}$, K is the nerve of the covering τ, and hence $|\text{nerv}\,\tau| \approx S^2$, and $|\text{nerv}\,{}^n\tau| \approx S^2$ for each $n \geq 0$. This implies that $K_A(S^2) \approx K_{\text{spec}}(S^2) \approx K_A(|\text{nerv}\,{}^n\tau|)$ for each $n \geq 0$.

We consider the directed set $\text{Cov}_f(X) \times \{|{}^n\tau|\}$ of coverings of $X \times S^2$, which is a cofinal subset of the set $\text{Cov}_f(X \times S^2)$. For each $(\alpha, {}^n\tau) \in \text{Cov}_f(X) \times \text{Cov}_f(S^2)$ the exterior K-multiplication is an isomorphism, since $|\text{nerv}\,{}^n\tau| \approx S^2$, hence the system $\{[p_{1\alpha}, p_2, {}^n\tau]\}_{(\alpha, {}^n\tau) \in \text{Cov}_f(X \times S^2)}$ of morphisms induces an isomorphism $K_{\text{spec}}(X) \times K_{\text{spec}}(S^2) \approx K_{\text{spec}}(X \times S^2)$.

Accordingly, we have

THEOREM 6. *The ring isomorphism*

$$K_{\text{spec}}(X) \times K_{\text{spec}}(S^2) \approx K_{\text{spec}}(X \times S^2)$$

is valid on the category of compact Hausdorff spaces.

This theorem gives us that the spectral K-functor can be defined also for positive n by setting $K_{\text{spec}}^{2n} = K_{\text{spec}}^{0}$ and $K_{\text{spec}}^{2n-1} = K_{\text{spec}}^{-1}$.

For each $\alpha \in \text{Cov}_f(X)$ let

$$\text{ch}_\alpha \colon K_1^{\#}(|\text{nerv}\,\alpha|) \to H^{\#}(|\text{nerv}\,\alpha|\,;\,Q)$$

be the Chern character. Then in view of the naturalness of the Chern character the system $\{\text{ch}_\alpha\}_{\alpha \in \text{Cov}_f(X)}$ of homomorphisms forms a morphism of direct spectra $\{K_A^{+}(|\text{nerv}\,\alpha|)\}_{\alpha \in \text{Cov}_f(X)} \to \{H^{\#}(|\text{nerv}\,\alpha|\,;\,Q)\}_{\alpha \in \text{Cov}_f(X)}$, and hence induces a homomorphism of limit groups

$$\text{ch}\colon K_{\text{spec}}^{\#}(X) \to H_{\text{spec}}^{\#}(X\,;\,Q),$$

where $H_{\text{spec}}^{\#}(X\,;\,Q)$ are the spectral cohomology groups of the space X, i.e., ch is a natural transformation of the spectral K-functor into the spectral cohomology functor.

For each $\alpha \in \text{Cov}_f(X)$ the homomorphism $\text{ch}_\alpha \times Q\colon K_A^{\#}(|\text{nerv}\,\alpha|) \times Q \to H^{\#}(|\text{nerv}\,\alpha|\,;\,Q)$ is an isomorphism, therefore, we have

Theorem 7. *On the category of compact Hausdorff spaces the homomorphism*

$$\text{ch} \times Q\colon K_{\text{spec}}^{\#}(X) \times Q \to H_{\text{spec}}^{\#}(X\,;\,Q)$$

is an isomorphism.

Bibliography

1. M. Atiyah and F. Hirzebruch, *Vector bundles and homogeneous spaces*, Proc. Sympos. Pure Math., vol. 3, Amer. Math. Soc., Providence, RI, 1961, pp. 7–38.
2. M. R. Bunyatov and S. A. Baĭramov, *K-theory on the category of distributive lattices*, Akad. Nauk Azerbaĭdzhan. SSR Dokl. **39** (1983), no. 5, 7–11; English transl. in this volume.
3. Edwin H. Spanier, *Algebraic topology*, McGraw-Hill, New York, 1966.
4. Samuel Eilenberg and Norman Steenrod, *Foundations of algebraic topology*, Princeton Univ. Press, Princeton, NJ, 1952.

Azerbaijan State University

Received 16/FEB/81

Translated by H. H. McFADEN

K-Theory on the Category of Boolean Algebras with Closure

UDC 513.831+513.836

M. R. BUNYATOV AND S. A. BAĬRAMOV

> Abstract. In this article vectorial bundles over closure Boolean algebras and K-theory on categories of closure Boolean algebras are introduced and investigated.

The K-functor was introduced by Atiyah and Hirzebruch. It was originally defined for algebraic varieties by Grothendieck. The construction of the K-functor was carried over to the category of smooth manifolds and CW-complexes in general by Atiyah and Hirzebruch [3], and Bukhshtaber and Mishchenko constructed a K-theory on the category of infinite cell complexes [8].

The K-theory constructed by Atiyah became a powerful method for investigating diverse problems in algebraic topology and functional analysis. The Bott periodicity theorem, the homotopy classifying space theorem for the K-functor, and the theorem on exact six-term sequences are the basic results of this theorem.

K-theory, together with cobordism theory, is an important extraordinary cohomology theory.

In the present article we introduce the concept of a vector bundle over a Boolean algebra with closure and use this concept to construct the K-functor on the category of Boolean algebras with closure. We define the Grothendieck cohomology and the induced homomorphisms for arbitrary Boolean algebras with closure and continuous complete Boolean homomorphisms. With this aim we introduce the category of cellular or polyhedral Boolean algebras with closure. For the K-functor on the category of polyhedral Boolean algebras

---

1991 *Mathematics Subject Classification.* Primary 55N15; Secondary 06E99.
Translation of Akad. Nauk Azerbaidzhan. SSR Dokl. 33 (1977), no. 12, 3–7.

with closure we prove an analogue of the homotopy classifying space theorem, the Bott isomorphism theorem, and the theorem on six-term exact sequences.

For each continuous mapping $f: X \to \gamma$ denote by f^* the covariant functor $f^*: \text{Vect}(\gamma) \to \text{Vect}(X)$ induced by f, which associates with each bundle $[E] \in \text{Vect}(\gamma)$ the induced bundle $[f^*(E)] \in \text{Vect}(X)$.

The rule $X \to \text{Vect}(X)$, $f \to f^*$ is a contravariant functor from the category Top of all topological spaces into the category of categories.

We associate with each inverse spectrum $X = (\{X_i\}_{i \in I}, \{p_i^{i'}\}_{i<i'})$ of topological spaces the direct spectrum

$$(\{\text{Vect}(X_i)\}_{i \in I}, \{(p_i^{i'})^*: \text{Vect}(X_i) \to \text{Vect}(X_{i'})\}_{i<i'})$$

of categories of vector bundles.

DEFINITION 1. The category $\text{Vect}(X) = \varinjlim \text{Vect}(X_i)$ is called the category of vector bundles over the inverse spectrum X.

The category $\text{Vect}(\Sigma(S))$ is called the category of vector bundles over the Boolean algebra with closure S ([1], [2]).

THEOREM 1. *The rule associating with each Boolean algebra with closure S the category $\text{Vect}(S)$ of vector bundles over the Boolean algebra S and with each continuous complete homomorphism $h: S \to S'$ of Boolean algebras with closure the induced morphism $h^{***}: \text{Vect}(S) \to \text{Vect}(S')$, where $h^*: \Sigma(S') \to \Sigma(S)$ is the induced morphism of inverse spectra and h^{***} the limit morphism induced by the morphism h^{**} of direct spectra of categories of vector bundles, is a covariant functor from the category of Boolean algebras with closure and their continuous complete homomorphisms into the category of categories of vector bundles over Boolean algebras.*

The functor $f^*: \text{Vect}(\gamma) \to \text{Vect}(X)$ of the induced bundle preserves tensor products and Whitney sums of vector bundles: $f^*(p \otimes q) = f^*(p) \otimes f^*(q)$ and $f^*(p \oplus q) = f^*(p) \oplus f^*(q)$ for any vector bundles $p: E \to \gamma$ and $q: F \to \gamma$. Hence, for each continuous mapping $f: X \to \gamma$ the induced covariant functor $f^*: \text{Vect}(\gamma) \to \text{Vect}(X)$ is a morphism of semirings.

Thus, the vector bundles $\text{Vect}(X)$ over an inverse spectrum X form a semiring with respect to the operations $\otimes$ and $\oplus$. In particular, the category $\text{Vect}(S)$ of vector bundles over a Boolean algebra with closure S is a $\otimes$-semiring.

DEFINITION 2. The ring $K[X] = K(\text{Vect}(X))$ that is the completion of the semiring $\text{Vect}(X)$ of vector bundles over the inverse spectrum X is called the Grothendieck ring of the inverse spectrum X. The ring $K(S) = K[\Sigma(S)]$ is called the Grothendieck ring of the Boolean algebra with closure S.

THEOREM 2. *The rule assigning to each Boolean algebra with closure S the Grothendieck ring $K(S)$ of S and to each continuous complete homomorphism $h: S \to S'$ of Boolean algebras the homomorphism $K(h): K(S) \to K(S')$ of Grothendieck rings induced by the morphism $h^{***}: \text{Vect}(S) \to$*

Vect(S') *of semirings is a covariant functor from the category of Boolean algebras and continuous complete homomorphisms of them into the category of Grothendieck rings.*

Let $X = (\{X_i\}_{i \in I}, \{p_i^{i'}\}_{i<i'})$ be an inverse spectrum of topological spaces, and $K(X_i)$ the Grothendieck ring that is the completion of the Vect(X_i) component of the direct spectrum

$$(\{\text{Vect}(X_i)\}_{i \in I}, \{(p_i^{i'})^*: \text{Vect}(X_i) \to \text{Vect}(X_{i'})\}_{i<i'})$$

of categories of vector bundles.

THEOREM 3. *The ring*

$$K(X) = (\{K(X_i)\}_{i \in I}, \{K((p_i^{i'})^*): K(X_i) \to K(X_{i'})\}_{i \in i'})$$

is a direct spectrum of rings. The limit $\varinjlim K(X)$ *is isomorphic to the Grothendieck ring* $K[X]$ *of the inverse spectrum* X. *Consequently, the limit* $\varinjlim \{K(\lambda)\}$ *of the direct spectrum*

$$(\{K(\lambda)\}_{\lambda \in \Sigma(S)}, \{K((p_\lambda^\mu)^*): K(\lambda) \to K(\mu)\}_{\lambda < \mu \in \Sigma(S)})$$

of rings is isomorphic to the Grothendieck ring of the Boolean algebra $K(S)$.

Let CW be the category of cell complexes, and Invspec(CW) the category of inverse spectra of cell complexes.

DEFINITION 3. A Boolean algebra with closure S together with an isomorphism $\kappa: S \xrightarrow{\sim} \varinjlim 2^X$ of the Boolean algebra S into the limit of a Boolean inverse spectrum of cell complexes $X \in \text{Invspec}(CW)$ is called a cellular Boolean algebra and denoted by the symbol (S, κ, X).

A morphism of a cellular Boolean algebra (S, κ, X) into a cellular Boolean algebra (S', κ, X') is a pair (h, f) formed from a continuous complete homomorphism $h: S \to S'$ and a morphism $f: X' \to X$ such that the following diagram is commutative:

$$\begin{array}{ccc} S & \xrightarrow{\kappa} & \varinjlim 2^X \\ h \downarrow & & \downarrow f^* \end{array}, \quad \kappa \cdot h = f^* \cdot \kappa.$$
$$\begin{array}{ccc} S' & \xrightarrow{\kappa} & \varinjlim 2^{X'} \end{array}$$

If for a Boolean algebra with closure S there is an inverse spectrum $X \in \text{Invspec}(CW)$ of cell complexes and an isomorphism $\kappa: S \xrightarrow{\sim} \varinjlim 2^X$, then one says that S admits a cell structure.

Obviously, the class of all cellular Boolean algebras and their morphisms forms a category with respect to the naturally defined composition. With each inverse spectrum $X \in \text{Invspec}(CW)$ of cell complexes we can associate the cellular Boolean algebra $\varinjlim 2^X$. This correspondence is a contravariant functor from the category Invspec(CW) to the category BoolCW of cellular Boolean algebras.

The category of cellular Boolean algebras with a distinguished element is introduced in the natural way. The objects of this category are triples (S, a, κ) formed from a Boolean algebra S, an isomorphism $\kappa\colon S \xrightarrow{\sim} \varinjlim 2^X$, and an element $a \in S$.

For each inverse spectrum $(X, \gamma) \in \operatorname{Invspec}(CW^2)$ of pairs of cell complexes there is a canonically associated cellular Boolean algebra with distinguished element $(\varinjlim 2^X, [\gamma])$, where $[\gamma] \stackrel{\text{def}}{=} [\gamma_i]$, $i \in I$, $[\gamma_i]$ being the equivalence class of $X_i \subset \varinjlim 2^X$ containing γ_i.

The Grothendieck ring for cellular Boolean algebras (S, κ, X) is defined by the formula $K[S, \kappa, X] \stackrel{\text{def}}{=} K[X]$.

We define the reduced $\widetilde{K}$-functor on the category of cellular Boolean algebras with distinguished element (S, a, κ). Suppose that $X \in \operatorname{Invspec}(CW_0)$ is an inverse spectrum $X = (\{(X_i, x_{0i})\}_{i \in I}, \{p_i^{i'} \colon (X_{i'}, x_{0i'}) \to (X_i, x_{0i})\}_{i < i'})$ of cell complexes with distinguished points, and $(\operatorname{id}_I, \operatorname{Imb})$ is a morphism of the inverse spectrum $\{x_{0i}\}_{i \in I}$ into the inverse spectrum $X = \{X_i\}_{i \in I}$ such that $\operatorname{Imb} = \{\operatorname{Imb}_i\}_{i \in I}$ and $\operatorname{Imb}_i(x_{0i}) = x_{0i}$. This morphism induces a morphism $(\operatorname{id}_i, K(\operatorname{Imb}^*))$ of the direct spectrum $\{K(X_i)\}_{i \in I}$ into the direct spectrum $\{K(x_{0i})\}_{i \in I}$, and all the rings $K(x_{0i})$ are isomorphic to the ring Z of integers.

We remark that the category $\operatorname{Dirspec}(I, \operatorname{Ring})$ of direct spectra of rings over a fixed index set I, which is a subcategory of the category $\operatorname{Dirspec}(\operatorname{Ring})$ of direct spectra of rings, is an Abelian category. In view of this, there exists a kernel of the morphism $K(\operatorname{Imb}^*)$ of direct spectra; denote it by $\widetilde{K}(X) = \operatorname{Ker} K(\operatorname{Imb}^*)$.

DEFINITION 4. *The reduced Grothendieck ring of a cellular Boolean algebra* $(\kappa; S, a)$, $a = [\{x_{0i}\}]$, *is defined to be the limit* $\varinjlim \widetilde{K}(x)$ *of the direct spectrum* $\widetilde{K}(x)$ *and denoted by* $\widetilde{K}[\kappa, S, a]$.

In the direct spectrum $\widetilde{K}(X) = \{\widetilde{K}(X_i)\}_{i \in I}$ the ring $\widetilde{K}(X_i)$ is the reduced ring over X_i for any i. Thus, the reduced Grothendieck ring $\widetilde{K}[\kappa; S, a]$ is the limit $\varinjlim \widetilde{K}(X)$ of the direct spectrum of the reduced rings of the vector bundles $\operatorname{Vect}(X_i)$ over X_i.

THEOREM 4. *The Grothendieck ring* $K[S, \kappa, X]$ *of the cellular Boolean algebra* (S, κ, X) *is isomorphic to the direct sum of the reduced ring* $\widetilde{K}[\kappa, S, a]$ *of the given Boolean algebra and the ring* Z *of integers*: $K[S, \kappa, X] \xrightarrow{\sim} \widetilde{K}[\kappa, S, a] \otimes Z$.

The bifunctor $\gamma^X\colon \operatorname{Top} \times \operatorname{Top} \to \operatorname{Ens}$, which is contravariant in X and covariant in γ, induces the bifunctor $C(X; \gamma)\colon \operatorname{Inv} \operatorname{Top} \times \operatorname{Top} \to \operatorname{Dir} \operatorname{Ens}$. In turn, the bifunctor $C(X; \gamma)$ induces the bifunctor $[X; \gamma]\colon \operatorname{Inv} \operatorname{Top} \times \operatorname{Top} \to \operatorname{Dir} \operatorname{Ens}$ on homotopy classes.

Thus, for any inverse spectrum $X = \{X_i\}_{i \in I}$ of connected finite cell complexes, $\{[X_i; B]\}_{i \in I}$ is a direct spectrum of groups, and for any i the group

$[X_i; B]$ is isomorphic to the group $\widetilde{K}(X_i)$ of the direct spectrum $\{\widetilde{K}(X_i)\}_{i \in I}$.

THEOREM 5. *On the category of cellular Boolean algebras the morphisms of the functors* $\Theta: \varinjlim[-; B] \to \widetilde{K}[-]$ *and* $\Theta: \varinjlim[-; B \times Z] \to K[-]$ *are isomorphisms of these functors, regarded as functors with values in the category of Abelian groups.*

The suspension functor S assigns to an inverse spectrum $X = \{(X_i, x_{0i})\}_{i \in I}$ of cell complexes with distinguished points the inverse spectrum $SX = \{SX_i\}_{i \in I}$ of suspensions. The cellular Boolean algebra corresponding to the inverse spectrum SX of suspensions is denoted by (S, κ, SX).

DEFINITION 5. For any integer $n \geq 0$ and any cellular Boolean algebra (S, κ, X) we set $\widetilde{K}^{-n}[S, \kappa, X] = \widetilde{K}[S, \kappa, S^n X]$. Similarly, for any cellular Boolean algebra with distinguished element $(\kappa; S, [\gamma])$ we set $K^{-n}[x; S, [\gamma]] = K[S, \kappa, S^n X \gamma]$ and $K^{-n}[S; \kappa, X] = \widetilde{K}[S, \kappa, S^n X^+]$.

Consider the direct spectra $\{K^{-(n+2)}(X_i, \gamma_i)\}_{i \in I}$ and $\{K^{-n}(X_i, \gamma_i)\}_{i \in I}$ of rings. As is known, for any i we have the Bott isomorphism $\beta: K^{-(n+2)}(X_i \gamma_i) \xrightarrow{\sim} K^{-n}(X_i, \gamma_i)$, therefore, the limits of these direct spectra are isomorphic: $K^{-(n+2)}[\kappa; S, [\gamma]] \xrightarrow{\sim} K^{-n}[\kappa; S, [\gamma]]$. This isomorphism is called the Bott isomorphism and denoted by β.

For any component $(X_i; \gamma_i)$ of the inverse spectrum $(X, \gamma) = \{(X_i, \gamma_i)\}_{i \in I}$ of pairs of cell complexes we have the exact sequence

$$K_\tau(X_i, \gamma_i) = \cdots \to K^{-1}(X_i, \gamma_i) \to K^{-1}(X_i) \to K^{-1}(\gamma_i) \to K^0(X_i, \gamma_i)$$
$$\to K^0(X_i) \to K^0(\gamma) \to K^1(X_i, \gamma_i) \to \cdots.$$

Thus, $\{K_\tau(X_i, \gamma_i)\}_{i \in I}$ is a direct spectrum of exact sequences, whose limit is also exact [7].

THEOREM 6. *The following sequences are exact for any cellular Boolean algebra with distinguished element* $[\kappa; S, [\gamma]]$:

a) $\cdots \to K^{-1}[\kappa; S, [\gamma]] \to K^{-1}[S, \kappa, X] \to K^{-1}[S, \kappa, \gamma] \to K^0[\kappa; S, [\gamma]] \to K^0[S, \kappa, X] \to K^0[S, \kappa, \gamma] \to \cdots$;

b) $\cdots \to \widetilde{K}^{-1}[\kappa; S, [\gamma]] \to \widetilde{K}^{-1}[S, \kappa, X] \to \widetilde{K}^{-1}[S, \kappa, \gamma] \to \widetilde{K}^0[\kappa; S, [\gamma]] \to \widetilde{K}^0[S, \kappa, X] \to \widetilde{K}^0[S, \kappa, \gamma] \to \cdots$.

BIBLIOGRAPHY

1. M. R. Bunyatov, *The Kolmogorov cohomology of abstract Boolean algebras with closure*, Dokl. Akad. Nauk SSSR **224** (1975), no. 1, 15–18; English transl. in Soviet Math. Dokl. **16** (1975).
2. _____, *Uniform Boolean algebras*, Dokl. Akad. Nauk SSSR **224** (1975), no. 2, 265–268; English transl. in Soviet Math. Dokl. **16** (1975).
3. M. Atiyah, *K-theory*, Lecture Notes, Harvard Univ., Cambridge, MA, 1964.
4. M. Atiyah and F. Hirzebruch, *Vector bundles and homogeneous spaces*, Proc. Sympos. Pure Math., vol. 3, Amer. Math. Soc., Providence, RI, 1961, pp. 7–38.
5. R. Bott, *Lectures on* $K(X)$, Mathematics Lecture Notes Series, Benjamin, New York and Amsterdam, 1969.

6. Dale Husemoller, *Fiber bundles*, McGraw-Hill, New York, 1966.
7. Samuel Eilenberg and Norman Steenrod, *Foundations of algebraic topology*, Princeton Univ. Press, Princeton, NJ, 1952.
8. V. M. Bukhshtaber and A. S. Mishchenko, *A K-theory on the category of infinite cell complexes*, Izv. Akad. Nauk SSSR Ser. Mat. **32** (1968), no. 3, 560–604; English transl. in Math. USSR-Izv. **2** (1968).

Azerbaijan State University

Received 30/JULY/77

Translated by H. H. McFADEN

On Index Theory for a Family of Fredholm Complexes

UDC 513.836

M. R. BUNYATOV AND S. A. BAĬRAMOV

Fredholm complexes were considered in [2], [7], [8], and [9] first and foremost in connection with the spectral theory of families of operators acting in a Hilbert space. Tensor products and the concept of an index were defined for Fredholm complexes, and the so-called stability of the Fredholm property for complexes was established.

First we introduce the space of Fredholm complexes, which enables us to make broader use of topological and function-theoretic properties of Fredholm complexes.

The concept of a family of Fredholm complexes on a topological space is a natural direct analogue of the concept of a Fredholm family on a topological space.

Since the space of Fredholm complexes is endowed with a topology, a homotopy classification of families of Fredholm complexes is possible. In contrast to the classical theory of families of Fredholm operators, the analogues of K-functors introduced below are not equipped with a group structure, due to the absence of an associative topological multiplication in $Fredholm(H, N)$. For families of Fredholm complexes it is no longer possible to introduce a concept of index that is a complete homotopy invariant. For example, if the space X consists of a single point, $X = \{*\}$, then even in this case the *index* mapping introduced above is not a complete homotopy invariant.

The basic objects to be investigated are Banach and Hilbert complexes, which are defined as follows.

Let $\{B_q\}_{q=0}^{N}$ be a family of Banach spaces, and $\{T_q\}_{q=0}^{N-1}$ a family of

---

1991 *Mathematics Subject Classification*. Primary 58B15; Secondary 18F25, 19K56, 46M20, 47A53, 58G12.

Translation of Problems in Geometry and Algebraic Topology, Azerbaĭdzhan. Gos. Univ., Baku, 1985, pp. 56–66.

bounded linear operators such that $T_q: B_q \to B_{q+1}$ for all $q = 0, \ldots, N-1$. If $T_{q+1} \cdot T_q = 0$ for all $q = 0, \ldots, N-1$, then one says that the pair $(\mathscr{B}, \mathscr{T}) = (\{B_q\}_{q=0}^N, \{T_q\}_{q=0}^{N-1})$ of families of Banach spaces and operators forms a complex of Banach spaces (here N can be finite or infinite). Further, one says that the sequence $\{T_q\}_{q=0}^{N-1}$ of operators forms a Banach complex over $\mathscr{B} = \{B_q\}_{q=0}^N$ and in this case $\mathscr{B}$ is called the base of this complex.

Denote by $Complex(\mathscr{B})$ the set of all Banach complexes over $\mathscr{B}$. This set has a natural topology. Indeed, denote by $\prod_{q=0}^{N-1} L(B_q, B_{q+1})$ for $N < \infty$ or by $\prod_{q=0}^{N} L(B_q, B_{q+1})$ for $N = \infty$ the product space, where $L(B_q, B_{q+1})$ is the space of bounded operators acting from B_q to B_{q+1}.

In what follows we consider mainly complexes with finite $N < \infty$. It is clear that the product $\prod_{q=0}^{N-1} L(B_q, B_{q+1})$ has a natural topological space structure. Then, since $Complex(\mathscr{B}) \subset \prod_{q=0}^{N-1} L(B_q, B_{q+1})$, it follows that $Complex(\mathscr{B})$ is a topological space in an obvious way; what is more, it is a metric space.

THEOREM 1. *$Complex(\mathscr{B})$ is a closed nowhere dense subspace of the space $\prod_{q=0}^{N-1} L(B_q, B_{q+1})$. The space $Complex(\mathscr{B})$ is arcwise connected.*

DEFINITION 1. The concatenation of complexes $T \in Complex(H, N)$ and $T' \in Complex(H, N')$ is defined to be the complex

$$T \odot T' \in Complex(H, N+N')$$

with the form

$$T \odot T' = H \xrightarrow{0} H \xrightarrow{T_1} H \to \cdots \xrightarrow{T_{N-1}} H \xrightarrow{0} H$$
$$\xrightarrow{\text{id}} H \xrightarrow{0} H \xrightarrow{T'_1} \cdots \xrightarrow{T'_{N'-1}} H \xrightarrow{0} H.$$

The next result can be proved directly.

THEOREM 2. 1) *A concatenation of exact (Fredholm) complexes is exact (Fredholm).*
 2) $(T \odot T')^* = T'^* \odot T^*$.
 3) $T \odot (T' \odot T'') = (T \odot T') \odot T''$.

Define $Fredholm_{\text{fin}}(H) = \bigcup_N Fredholm(H, N)$.

THEOREM 3. 1. *$Fredholm_{\text{fin}}(H)$ is a semigroup with involution with respect to the concatenation operation. The mapping* $\text{index}: Fredholm_{\text{fin}}(H) \to \mathbf{Z}$, *assigning to each complex T its index* $\text{index}(T)$ *is surjective.*
 2. $\text{index}(T \odot T') = \text{index}(T) + (-1)^T \text{index}(T')$.
 3. $\text{index}(T^*) = (-1)^{|T|} \text{index}(T)$, *where $|T|$ is the length of T.*

THEOREM 4. *The mapping* $\text{index}: Fredholm(H, N) \to \mathbf{Z}$ *is a continuous mapping, hence it is constant on each connected component of the space $Fredholm(H, N)$.*

THEOREM 5. *Fredholm(H, N) is open in the space Complex(H, N) of complexes.*

This follows immediately from the stability theorem in [9].

DEFINITION 2. A continuous mapping $f: X \to Fredholm(H, N)$ is called a family of Fredholm complexes.

In this way the spaces $Fredholm(H, N)$ can be used to construct new homotopy invariants of topological spaces on the one hand, and to investigate families of Fredholm complexes in the large on the other hand.

DEFINITION 3. The set $[X, Fredholm(H, N)]$ of homotopy classes of continuous mappings from the space X to the space $Fredholm(H, N)$ is called the functor of homotopy types of families of Fredholm complexes on X, and denoted by $K^N(X)$.

Thus, we have a bifunctor $K^N(X)$ that is contravariant in the first argument and covariant in the second.

Here we consider at the same time different continuous families $\delta: X \to Complex(K_0, \ldots, K_N)$ and $\partial: Y \to Complex(L_0, \ldots, L_N)$ with different $(K_0, \ldots, K_N)$ and $(L_0, \ldots, L_N)$.

We introduce the category of families of complexes. Since the specification of a complex is equivalent to the specification of a differential on the corresponding graded object formed by the members of the complex, instead of a family of complexes over a given graded Banach space $E = \bigoplus_{q=0}^{N} E_q$ we sometimes speak of the family of corresponding differentials of these complexes.

Let $\delta: X \to \mathrm{Der}(E_0, \ldots, E_N)$ and $\partial: Y \to \mathrm{Der}(F_0, \ldots, F_N)$ be two families of derivations.

A morphism from the family δ to the family ∂ is defined to be a pair (f, φ) formed from a continuous mapping $f: X \to Y$ and a family $\varphi = \{\varphi_x\}_{x \in X}$ of chain transformations $\varphi_x: \delta_x \to \partial_{f(x)}$ satisfying the condition that the family $\varphi: X \to \prod_{q=0}^{N} \mathrm{Hom}(E_q, F_q)$ is a continuous mapping.

THEOREM 6. *The class of all families of complexes and their morphisms forms a category. Denote this category by Familycomplex.*

We consider the subcategory of *Familycomplex* consisting of the families of complexes over one and the same space X; denote this subcategory by $Familycomplex_X$.

In the category $Familycomplex_X$ we introduce the relation of being chain homotopic. A chain homotopy of morphisms (id, φ_0) and (id, φ_1) is defined to be a continuous mapping $\mathscr{H}: X \to \prod_{q=1}^{N} \mathrm{Hom}(E_q, F_{q-1})$ such that $\mathscr{H}(x)$ is a chain homotopy from φ_{0x} to φ_{1x} for each $x \in X$.

THEOREM 7. *The relation of being chain homotopic is an equivalence relation in the class of morphisms from one family to another, and it is natural with respect to composition.*

Let $\delta\colon X \to Fredholm(E_0, \ldots, E_N)$ be a family of Fredholm complexes over Hilbert spaces $E_0, \ldots, E_N$. Then since $\operatorname{Im}\delta_x^{q-1} \subset \ker\delta_x^q \subset E_q$, it follows that $H^q(\delta_x) = \ker\delta_x^q/\operatorname{Im}\delta_x^{q-1}$ can be canonically identified with the orthogonal complement of $\operatorname{Im}\delta_x^{q-1}$ in $\ker\delta_x^q$, and hence $H^q(\delta_x)$ can be identified in a natural way with a subspace of E_q. In view of this we can form the subspace

$$\mathscr{H}^q(\delta) = \{(x, u) | x \in X,\ u \in H^q(\delta_x)\},$$

considered together with the projection on the first component $\operatorname{pr}_1 \colon \mathscr{H}^q(\delta) \to X$; $\mathscr{H}^q(\delta)$ forms a finite-dimensional vector quasibundle, which we also denote by $\mathscr{H}^q(\delta)$ and call a homology vector quasibundle.

Suppose now that $(f, \varphi)\colon \delta \to \partial$ is a morphism of two families. We define a mapping $(f, \varphi)_{xq}\colon \mathscr{H}^q(\delta) \to \mathscr{H}^q(\partial)$ by the formula

$$(f, \varphi)_{xq}(x, u) = (f(x), \varphi_{xq}(u)) \quad \text{for } (x, u) \in \mathscr{H}^q(\delta).$$

Note that $(f, \varphi)_{xq}$ is a linear operator on each fiber. Thus, we have a morphism of vector quasibundles:

$$(f, \varphi)_{xq}\colon \mathscr{H}^q(\delta) \to \mathscr{H}^q(\partial).$$

THEOREM 8. *The rule associating with each family*

$$\delta\colon X \to Fredholm(E_0, \ldots, E_N)$$

of Fredholm complexes its homology quasibundle $\mathscr{H}^q(\delta)$ and to each morphism $(f, \varphi)\colon \delta \to \partial$ of Fredholm families the morphism $(f, \varphi)_{xq}\colon \mathscr{H}^q(\delta) \to \mathscr{H}^q(\partial)$ is a covariant functor from the category FamilyFredholm of families of Fredholm complexes into the category Quasivect of finite-dimensional vector quasibundles.

In view of the chain homotopy invariance of the cohomology groups of chain complexes we have

THEOREM 9. *The homotopy vector quasibundle of families of Fredholm complexes is a homotopy invariant.*

We now use the functor of continuous sections of vector quasibundles. For a vector quasibundle $E \to X$ over a topological space X denote by $\Gamma(E)$ the $C(X)$-module formed by all continuous sections of this bundle. If X is compact, then the ring $C(X)$ is naturally endowed with the structure of a Banach algebra, and hence $\Gamma(E)$ is a module over the commutative Banach algebra $C(X)$. Accordingly, we have the functor

$$\Gamma^q = \Gamma \cdot \mathscr{H}^q \colon FamilyFredholm_N(X) \to Modul_{C(X)}.$$

DEFINITION 4. Let $\delta\colon X \to Fredholm(E_0, \ldots, E_N)$ be a family of Fredholm complexes. For any $q = 0, \ldots, N$ we call $\Gamma(\mathscr{H}^q(\delta))$ the q-dimensional cohomology module of the family δ and denote it by the symbol $\Gamma^q(\delta)$.

By virtue of the foregoing the relation $\delta \mapsto \Gamma^q(\delta)$, $(f, \varphi): \delta \to \partial \mapsto (f, \varphi)^*: \Gamma^q(\delta) \to \Gamma^q(\partial)$ is a contravariant functor from the category *FamilyFredholm$_N$* to the category of modules over Banach algebras.

Suppose now that $f: X \to Y$ is a homeomorphism. We define an operator $\Gamma(f, \varphi): \Gamma(\mathscr{H}^q(\delta)) \to \Gamma(\mathscr{H}^q(\partial))$ by the formula $\Gamma(f, \varphi)(s)(y) = \varphi_{0q}(s(f^{-1}(y)))$ for any section $s \in \Gamma(\mathscr{H}^q(\delta))$ and any $y \in Y$. In other words, $\Gamma(f, \varphi)(s) = \varphi_{0q} \cdot s \cdot f^{-1}$. It is clear from the definition that $\Gamma(f, \varphi)(s) \in \Gamma(\mathscr{H}^q(\partial))$. Obviously, $\Gamma(f, \varphi)$ is a homomorphism.

We consider also the ring homomorphism $(f^{-1})^*: C(X) \to C(Y)$. The pair $((f^{-1})^*, \Gamma(f, \varphi))$ is a morphism

$$(C(X), \Gamma(\mathscr{H}^q(\delta))) \to (C(Y), \Gamma(\mathscr{H}^q(\partial)))$$

of modules.

The morphism (f, φ) together with the homotopy $f: X \to Y$ is called a morphism with change of variable of the parameter of the family. The subcategory formed by such morphisms of the category *FamilyFredholm$_N$* is denoted by *FamilyFredholm$_N^s$*.

THEOREM 10. *The relation*

$$\delta \mapsto \Gamma^q(\delta), \quad (f, \varphi): \delta \to \partial \mapsto \Gamma(f, \varphi): \Gamma^q(\delta) \to \Gamma^q(\partial)$$

is a covariant functor from the category FamilyFredholm$_N^s$ to the category $\mathscr{MB}$.

The homology theory obtained with $\Gamma^q(\delta)$ satisfies all the axioms for a homology theory, and the groups $\Gamma^q(\delta)$ are homotopy invariants.

DEFINITION 5. A sequence $\delta \xrightarrow{\varphi} \partial \xrightarrow{\psi} \Delta$ of families of Fredholm complexes is said to be exact if for all $x \in X$ the sequence $\delta_x \xrightarrow{\varphi_x} \partial_x \xrightarrow{\psi_x} \Delta_x$ is exact.

Suppose that $0 \to \delta \xrightarrow{\varphi} \partial \xrightarrow{\psi} \Delta \to 0$ is an exact sequence, and consider for each $x \in X$ the boundary $\Gamma_x: H^q(\Delta_x) \to H^{q-1}(\delta_x)$ generated by the short exact sequence $0 \to \delta_x \xrightarrow{\varphi_x} \partial_x \xrightarrow{\psi_x} \Delta_x \to 0$ of complexes.

We consider the mapping $X \to L(E_q, E'_{q+1})$ assigning to each $x \in X$ the operator $\overline{\Lambda}_x = i_{H^q(\delta_x)} \circ \Lambda_x \circ \mathrm{pr}_{H^q(\Delta_x)}$. The short exact sequence $0 \to \delta \xrightarrow{\varphi} \partial \xrightarrow{\psi} \Delta \to 0$ is said to be regular if the mapping $\overline{\Lambda}: X \to L(E_q, E'_{q+1})$ is continuous. In this case $\overline{\Lambda}$ induces a morphism $\overline{\Lambda}_*: \mathscr{H}^q(\Delta) \to \mathscr{H}^{q+1}(\delta)$ of quasibundles, which is also assumed to be continuous.

THEOREM 11. *For any regular short exact sequence $0 \to \delta \to \partial \to \Delta \to 0$ of families of Fredholm complexes the sequence*

$$\cdots \to \mathscr{H}^q(\delta) \to \mathscr{H}^q(\partial) \to \mathscr{H}^q(\Delta) \to \mathscr{H}^{q+1}(\delta) \to \cdots$$

of vector quasibundles is exact.

THEOREM 12. *Suppose that $0 \to \delta \xrightarrow{\varphi} \partial \xrightarrow{\psi} \Delta \to 0$ is an arbitrary regular short exact sequence of families of Fredholm complexes. For arbitrary* sections s_1, s_2, and s_3 of the respective quasibundles $\mathscr{H}^q(\partial_x)$, $\mathscr{H}^q(\Delta_x)$, and $\mathscr{H}^{q+1}(\delta_x)$ we write $L_{s_1(x)} = P_{q*x}^{-1}(s_1(x))$, $L_{s_2(x)} = \psi_{q*x}^{-1}(s_2(x))$, and $L_{s_3(x)} = \Lambda_{*qx}^{-1}(s_3(x))$, and we form the subbundles

$$L_{s_i} = \{(x, u) | x \in X, u \in L_{s_i(x)}\}, \quad i = 1, 2, 3,$$

and assume that the subbundles L_{s_i} have continuous sections; then the sequence

$$\cdots \to \Gamma^q(\delta) \xrightarrow{\Gamma^q(\varphi)} \Gamma^q(\partial) \xrightarrow{\Gamma^q(\psi)} \Gamma^q(\Delta) \xrightarrow{\overline{\Lambda}_*} \Gamma^{q+1}(\delta) \to \cdots,$$

called the homology sequence for the short exact sequence $0 \to \delta \xrightarrow{\varphi} \partial \xrightarrow{\psi} \Delta \to 0$, is exact.

Using these homology invariants, we can introduce the concept of the index of a family of Fredholm complexes in a natural way. With this aim we take the Grothendieck group $K(Modul_{C(X)})$ of the Abelian category of modules over the ring $C(X)$ as the group of values of the index [5].

Denote by $[A]$ the element in $K(Modul_{C(X)})$ containing the module A as a representative; in other words, $[A]$ is the canonical imbedding of the monoid $Modul_{C(X)}$ with respect to the direct sum in the Grothendieck group $K(Modul_{C(X)})$ of $Modul_{C(X)}$.

DEFINITION 6. The index of the family of Fredholm complexes $\delta: X \to Fredholm_N$ is defined to be the element

$$\text{Index}(\delta) = \sum_{q=0}^{N} (-1)^q [\Gamma^q(\delta)]$$

of the group $K(Modul_{C(X)})$.

THEOREM 13. *Suppose that $0 \to \delta \to \partial \to \Delta \to 0$ is a regular short exact sequence of families of Fredholm complexes, and assume the condition of Theorem 12. Then*

$$\text{Index}\,\partial = \text{Index}\,\delta + \text{Index}\,\Delta.$$

THEOREM 14. *Chain-homotopy equivalent families of Fredholm complexes have equal indices.*

THEOREM 15. *If a space X consists of a single point, then, for any family $\delta: X \to Fredholm(E_0, \ldots, E_N)$, $\text{Index}(\delta) = \text{index}(\delta_x)$, where δ_x is a Fredholm complex.*

We now consider the question of stability of a homology group for families of Fredholm complexes.

Let $\delta: X \to \text{Der}(E_0, \ldots, E_N)$ be a holomorphic function on a complex space X with values in the set of derivations over a graded Hilbert space $E = \bigoplus_{q=0}^{N} E_q$. Suppose that there is an $x_0 \in X$ such that δ_{x_0} splits. Then at some point x of a neighborhood $U \ni x_0$ the family $\delta|_U$ is holomorphically structurally chain-homotopically equivalent to some family of derivations $\partial\colon U \to \mathrm{Der}(H^0(\delta_{x_0}), \ldots, H^N(\delta_{x_0}))$ in the graded homology group of the complex δ_{x_0}.

DEFINITION 7. A complex $\delta_0 \in Complex(E_0, \ldots, E_N)$ is said to be structurally holomorphically homologically stable if for any holomorphic family $\delta\colon X \to \mathrm{Der}(E_0, \ldots, E_N)$ containing the complex δ_0, i.e., such that $\delta(x_0) = \delta_0$ for some $x_0 \in X$, there is a neighborhood U of x_0 such that $\delta|_U$ is holomorphically chain-homotopically equivalent to some family $H(\delta)\colon U \to \mathrm{Der}(H^0(\delta_0), \ldots, H^N(\delta_0))$.

PROPOSITION 2. *Any holomorphic family of complexes is locally Fredholm at a point $x_0 \in X$ satisfying the Fredholm property condition, i.e., for any holomorphic family of complexes $\delta\colon X \to \mathrm{Der}(E_0, \ldots, E_N)$ and for any point $x_0 \in X$ with the complex $\delta(x_0)$ Fredholm the family $\delta|_U$ is a holomorphic family of Fredholm complexes.*

THEOREM 16. *Suppose that $\delta\colon X \to \mathrm{Der}(E_0, \ldots, E_N)$ is a holomorphic family of derivations in a graded Hilbert space $E = E_1 \oplus \cdots \oplus E_N$, and suppose that at some point $x_0 \in X$ the complex $\delta(x_0)$ is homologically structurally stable and the sequences $0 \to H^q(\delta_{x_0}) \to F^q$ are exact for all $q = 0, \ldots, N$. Then there is a $U \in N(x_0)$ such that the sequence $0 \to H^q(\delta(x)) \to F^q$ is exact for all $x \in U$.*

BIBLIOGRAPHY

1. M. F. Atiyah, *K-theory*, Benjamin, New York and Amsterdam, 1967.
2. A. S. Dynin, *K-theory and pseudodifferential operators*, Seventh Summer Mathematical School (Kaciveli, 1969), Izdanie Mat. Akad. Nauk Ukrain. SSR, Kiev, 1970, pp. 144–190. (Russian)
3. G. A. Isaev and A. S. Faĭnshteĭn, *Joint spectra of finite commutative families*, Spectral Theory of Operators, No. 3, "Èlm", Baku, 1980, pp. 222–257. (Russian)
4. Max Karoubi and Orlando Villamayor, *K-theorie algebrique et K-theorie topologique. I*, Math. Scand. **28** (1971), 265–307.
5. Serge Lang, *Algebra*, Addison-Wesley, Reading, MA, 1965.
6. V. P. Palamodov, *Deformation of complex spaces*, Uspekhi Mat. Nauk **31** (1976), no. 3, 129–194; English transl. in Russian Math. Surveys **31** (1976).
7. M. Schechter and M. Snow, *The Fredholm spectrum on tensor products*, Proc. Roy. Irish Acad. Sect. A **75** (1975), no. 13, 121–127.
8. Joseph L. Taylor, *A joint spectrum for several commuting operators*, J. Funct. Anal. **6** (1970), no. 2, 172–191.
9. A. S. Faĭnshteĭn, *Joint essential spectrum of a family of linear operators*, Funktsional. Anal. i Prilozhen. **14** (1980), no. 2, 83–84; English transl. in Functional Anal. Appl. **14** (1980).
10. Paul R. Halmos, *A Hilbert space problem book*, Van Nostrand, Princeton, NJ, 1967.

Translated by H. H. McFADEN

On Noncommutative Function Spaces

A. M. BIKCHENTAEV

§1. Terminology and notation

Let X be a metrizable topological vector space (MTVS) over the field $\mathbb{C}$, and let Θ be the set of all complex-valued null sequences in $\mathbb{C}$.

A map $\|\cdot\|\colon X \to [0,\infty)$ is called an F-norm if the following conditions hold:

(F1) $\|x\| = 0$ if and only if $x = 0$;
(F2) $\|ax\| = \|x\|$ for all $x \in X$ and $a \in \mathbb{C}$, $|a| = 1$;
(F3) $\|x + y\| \leq \|x\| + \|y\|$ for all $x, y \in X$;
(F4) if $(a_n) \in \Theta$, then $(\|a_n x\|) \in \Theta$ for all $x \in X$.

A set $Y \subset X$ is said to be bounded if for each neighborhood U of the origin there exists $\lambda > 0$ such that $Y \subset \lambda U$. In other words, $(\|a_n x_n\|) \in \Theta$ for all sequences (x_n) in Y and $(a_n) \in \Theta$. A TVS containing a bounded neighborhood of the origin is said to be locally bounded. On a locally bounded MTVS X there exists a p-homogeneous F-norm (for some $p \in (0, 1]$) giving the original topology [4]. Here by a p-homogeneous norm we mean a norm such that $\|ax\| = |a|^p \cdot \|x\|$ for all $x \in X$ and $a \in \mathbb{C}$.

In what follows we suppose $\mathscr{M}$ to be a fixed von Neumann algebra with a faithful normal semifinite trace τ; $\mathscr{M}^{\mathrm{P}}$ and $\mathscr{M}^{\mathrm{U}}$ are the sets of projections and unitary operators in $\mathscr{M}$, respectively; $\|\cdot\|_\infty$ is the usual operator norm on $\mathscr{M}$. We denote by $\{E_\lambda^A, \lambda \in \mathbb{R}\}$ the resolution of the identity corresponding to a selfadjoint operator A and by E_{p} the largest projection in $\mathscr{M} \cap \mathscr{M}'$ ($\mathscr{M}'$ is the commutant of $\mathscr{M}$) for which $\mathscr{M}_{E_{\mathrm{p}}}$ is of type P.

In [5] the set $\mathscr{K}$ of totally measurable operators (with respect to τ) was introduced; this is the set of closed densely defined operators A affiliated with $\mathscr{M}$ and satisfying the condition: $\tau(\mathbf{1} - E_\lambda^{|A|}) < \infty$ for some $\lambda > 0$. It

---

1991 *Mathematics Subject Classification.* Primary 46L10, 47D25; Secondary 46A32, 47D15, 47D40.

Translation of Functional Analysis. Spectral Theory, Interinstitutional collection of scientific works, Ul′yanovsk. Gos. Ped. Inst., Ul′yanovsk, 1987, pp. 33–43.

is known [5] that $\mathscr{K}$ is closed with respect to the operations of conjugation, multiplication by a scalar, strong convergence, and multiplication of operators; $\mathscr{K}$ is a complete MTVS with respect to the topology t_τ defined by the fundamental system of neighborhoods of the origin

$$\mathscr{U}(\varepsilon, \delta) = \{A \in \mathscr{K} \mid \exists E \in \mathscr{M}^{\mathrm{P}} : \|AE\|_\infty \leq \varepsilon, \tau(\mathbf{1} - E) \leq \delta\}$$

and called the topology of convergence in measure. As usual, L_1 is the Banach space of integrable operators (with respect to τ) (see, for example, [3], [5]).

DEFINITION 1.1. By a nonincreasing rearrangement $A^\sim$ of an operator $A \in \mathscr{K}$ we mean the function $A^\sim : (0, \infty) \to [0, \infty)$ defined by the equality

$$A^\sim(\alpha) = \inf\{\lambda > 0 \mid \tau(\mathbf{1} - E_\lambda^{|A|}) \leq \alpha\}.$$

In the next proposition we collect together facts concerning nonincreasing rearrangements given in [5] and some trivial consequences of them.

PROPOSITION 1.2. (i) $A^\sim(\alpha) = \inf\{\|AE\|_\infty \mid E \in \mathscr{M}^{\mathrm{P}}, \tau(\mathbf{1} - E) \leq \alpha\}$.
(ii) $A^{*\sim} = |A|^\sim = A^\sim = (U^*AU)^\sim$ $(A \in \mathscr{K}, U \in \mathscr{M}^{\mathrm{U}})$.
(iii) $(AB)^\sim \leq \|B\|_\infty \cdot A^\sim$, $(BA)^\sim \leq \|B\|_\infty \cdot A^\sim$ $(A \in \mathscr{K}, B \in \mathscr{M})$.
(iv) Let $A, B \in \mathscr{K}$ and $\alpha, \beta > 0$. Then
  (a) $(A + B)^\sim(\alpha + \beta) \leq A^\sim(\alpha) + B^\sim(\beta)$;
  (b) $(AB)^\sim(\alpha + \beta) \leq A^\sim(\alpha) \cdot B^\sim(\beta)$.
(v) $A^\sim \leq B^\sim$ if $0 \leq A \leq B$ $(A, B \in \mathscr{K})$.
(vi) $A \in \mathscr{M}$ if and only if $A \in \mathscr{K}$ and $A^\sim(0+) = \lim_{\alpha \searrow 0} A^\sim(\alpha) = \sup_{\alpha > 0} A^\sim(\alpha) < \infty$; furthermore, $\|A\|_\infty = A^\sim(0+)$.
(vii) A sequence of operators (A_n) in $\mathscr{K}$ converges in measure to the operator $A \in \mathscr{K}$ if and only if $((A_n - A)^\sim(\alpha)) \in \Theta$ for any $\alpha > 0$.

§2. The spaces $L_f(\mathscr{M}, \tau)$ associated with a faithful normal semifinite trace τ on a von Neumann algebra $\mathscr{M}$

Let Φ_r, $r \in \{1, \infty\}$, be the class of continuous strictly increasing functions f defined on $[0, r)$ and satisfying the condition $f(0) = 0$. In what follows, $f \in \Phi_1 \cup \Phi_\infty$ unless otherwise stipulated.

Let $C(\mathscr{M})$ be the algebra of measurable operators affiliated with $\mathscr{M}$ ([3], Definition 2.1).

PROPOSITION 2.1. $L_f = L_f(\mathscr{M}, \tau) = \{A \in C(\mathscr{M}) \mid \exists l_A > 0 : f(l_A|A|) \in L_1\}$ is a $*$-lineal in $\mathscr{K}$.

PROOF. Let $f \in \Phi_\infty$. Since for $A \in C(\mathscr{M})$

$$f(|A|) = \int_0^\infty f(\lambda)\,dE_\lambda^{|A|} = \int_0^\infty s\,dE_{s'}^{f(|A|)},$$

where

$$E_{s'}^{f(|A|)} = \begin{cases} E_{f^{-1}(s)}^{|A|} & \text{if } s \in [0, \lim_{\lambda \to \infty} f(\lambda)), \\ 1 & \text{if } s \notin [0, \lim_{\lambda \to \infty} f(\lambda)) \end{cases}$$

([2], p. 367), it is clear that $L_f \subset \mathcal{H}$. For negative $A \in \mathcal{H}$ we have $f(A)^\sim = f(A^\sim)$. In fact,

$$f(A)^\sim(\alpha) = \inf\{s > 0 \mid \tau(1 - E_{s'}^{f(A)}) \leq \alpha\} = \inf\{s > 0 \mid \tau(1 - E_{f^{-1}(s)}^A) \leq \alpha\}$$
$$= \inf\{f(\lambda) \mid \lambda > 0, \tau(1 - E_\lambda^A) \leq \alpha\} = f(\inf\{\lambda > 0 \mid \tau(1 - E_\lambda^A) \leq \alpha\})$$
$$= f(A^\sim(\alpha)).$$

Let $A \in L_f$ and $l_A > 0$ satisfy the condition

$$\tau(f(l_A \cdot |A|)) = \int_0^\infty f(l_A \cdot A^\sim(\alpha))\, d\alpha < \infty.$$

It follows from (ii) of Proposition 1.2 that $|A|$, $A^* \in L_f$. Suppose further that $B \in L_f$, $2l = \min\{l_A, l_B\}$, and $\mathbb{R}' \equiv \{\alpha > 0 \mid A^\sim(\frac{\alpha}{2}) \geq B^\sim(\frac{\alpha}{2})\}$. Then

$$\tau(f(l \cdot |A+B|)) = \int_0^\infty f(l(A+B)^\sim(\alpha))\, d\alpha$$
$$\leq \int_0^\infty f\left(l \cdot A^\sim\left(\frac{\alpha}{2}\right) + l \cdot B^\sim\left(\frac{\alpha}{2}\right)\right) d\alpha$$
$$\leq \int_{\mathbb{R}'} f\left(2l \cdot A^\sim\left(\frac{\alpha}{2}\right)\right) d\alpha + \int_{(0,\infty)\setminus\mathbb{R}'} f\left(2l \cdot B^\sim\left(\frac{\alpha}{2}\right)\right) d\alpha$$
$$\leq 2(\tau(f(l_A \cdot |A|)) + \tau(f(l_B|B|))) < \infty,$$

so that $A+B \in L_f$. If $a \in \mathbb{C}\setminus\{0\}$, then $\tau(f(l_A \cdot |A|)) = \tau(f(|a|^{-1}l_A|aA|)) < \infty$ and $aA \in L_f$.

Suppose now that $f \in \Phi_1$. Then $l_A > 0$ for $A \in L_f$ is chosen from the condition $E_{1-0}^{l_A \cdot |A|} = 1$ and $L_f \subset \mathcal{M}$. If $A = V \cdot |A|$ is the polar decomposition, then it follows from the relation $E_\lambda^{|A^*|} = VE_\lambda^{|A|}V^* + (1 - VV^*)$ that $E_{1-0}^{l_A|A^*|} = 1$ as well. The choice of the quantity $2l = \min\{l_A, l_B\}$ for $A, B \in L_f$ gives $E_{1-0}^{l|A+B|} = 1$. Later on we shall, where possible, omit from the proofs certain details relating to the case $r = 1$.

PROPOSITION 2.2. *The map* $\|\cdot\|_f \colon L_f \to [0, \infty)$ *defined by the equality*

$$\|A\|_f = \inf\{\varepsilon > 0 \mid \tau(f(\varepsilon^{-1} \cdot |A|)) < \varepsilon\}$$

is an F-norm on L_f.

PROOF. Conditions (F1) and (F2) clearly hold. We will prove (F3). Let $A, B \in L_f$ and $\varepsilon > 0$ be arbitrary. By the definition of $\|\cdot\|_f$, there exist $a, b > 0$ such that $\tau(f(a^{-1}|A|)) < a$, $\tau(f(b^{-1}|B|)) < b$ and $\|A\|_f < a <$

(ii) $f \prec g$ *if and only if* $g^{-1} \prec f^{-1}$, *where* f^{-1} *is the inverse of* f. *Hence*, $f \asymp g$ *if and only if* $f^{-1} \asymp g^{-1}$.

LEMMA 2.9. (i) *If*
$$f, g, \varphi \in \Phi_r \quad \text{and} \quad f^{-1} \cdot g^{-1} \prec \varphi^{-1}, \qquad (*)$$
then $L_f L_g \subset L_\varphi$ *and we have the following:*
  (a) $\|AB\|_\varphi \le C \max\{\|A\|_f \cdot \|B\|_g, \|A_f\| + \|B\|_g\}$ ($A \in L_f$, $B \in L_g$, *and* C *is the constant involved in the definition of* $\prec$ *in* $(*)$);
  (b) $\forall A \in L_f \; \forall \varepsilon > 0 \; \exists \delta > 0 : (B \in L_g, \|B\|_g < \delta \Rightarrow \|AB\|_\varphi < \varepsilon)$;
  (c) *the map* $(A, B) \to AB$ $(\langle L_f, \|\cdot\|_f\rangle \times \langle L_g, \|\cdot\|_g\rangle \to \langle L_\varphi, \|\cdot\|_\varphi\rangle)$ *is continuous*.

(ii) *If* $f, g, \varphi \in \Phi_r$ *and* $f^{-1} \cdot g^{-1} \asymp \varphi^{-1}$, *then* $L_f L_g = L_\varphi$.

PROOF. (i)(a) Let $A \in L_f$, $B \in L_g$, $\varepsilon > 0$ be arbitrary, C be the constant involved in the definition of $\prec$ in $(*)$, and $a, b > 0$ be such that
$$\|A\|_f < aC^{-1/2} < \|A\|_f + \varepsilon, \qquad \|B\|_g < BC^{-1/2} < \|B\|_g + \varepsilon.$$
Then we have:
$$\tau(1 - E_{C\lambda}^{\varphi((ab)^{-1}|AB|)}) = \tau(1 - E_{ab\varphi^{-1}(C\lambda)}^{|AB|}) \le \tau(1 - E_{abC^{-1}f^{-1}(\lambda)g^{-1}(\lambda)}^{|AB|})$$
$$\le \tau(1 - E_{aC^{-1/2}f^{-1}(\lambda)}^{|A|}) + \tau(1 - E_{bC^{-1/2}g^{-1}(\lambda)}^{|B|})$$
$$= \tau(1 - E_\lambda^{f(a^{-1}C^{1/2}|A|)}) + \tau(1 - E_\lambda^{g(b^{-1}C^{1/2}|B|)}),$$
$$C^{-1} \int_0^\infty \tau(1 - E_\lambda^{\varphi((ab)^{-1}|AB|)}) d\lambda \le \int_0^\infty \tau(1 - E_\lambda^{f(a^{-1}C^{1/2}|A|)}) d\lambda$$
$$+ \int_0^\infty \tau(1 - E_\lambda^{g(b^{-1}C^{1/2}|B|)}) d\lambda$$
$$< (a+b)C^{-1/2}.$$

Therefore,
$$\tau(\varphi(\max\{ab, C^{1/2}(a+b)\}^{-1} \cdot |AB|))$$
$$\le \tau(\varphi((ab)^{-1} \cdot |AB|)) < \max\{ab, C^{1/2}(a+b)\},$$
$AB \in L_f$, and (a) is obtained. To verify (b) we prove the inequality
$$\varphi(st) \le C \max\{f(C^{1/2}s), g(C^{1/2}t)\} \quad \text{for } 0 < s, t < C^{-1/2}r.$$
We set
$$\lambda_1 = f^{-1}(C^{-1}\varphi(st)), \qquad \lambda_2 = g^{-1}(C^{-1}\varphi(st)).$$
If $\lambda_1 \le C^{1/2}s$, then $f(\lambda_1) = C^{-1}\varphi(st) \le f(C^{1/2}s)$. Since $\lambda_1 \cdot \lambda_2 = Cst$, it follows from $\lambda_1 > C^{1/2}s$ that $\lambda_2 < C^{1/2}t$ and $C^{-1}\varphi(st) = g(\lambda_2) < g(C^{1/2}t)$. Now let $A \in L_f$, $\varepsilon > 0$ be arbitrary, $\beta \equiv \varepsilon/4C$, and let $\delta \in (0, \beta)$ be chosen from the condition $\tau(f(C\delta\varepsilon^{-1}|A|)) < \beta$. It follows from (iv)(b) of

Proposition 1.2 and from the fact, proved above, that $\|B\|_g < \delta$ for $B \in L_g$ that:

$$\int_0^\infty \varphi(\varepsilon^{-1}(AB)^\sim(\alpha))\,d\alpha \le 2\int_0^\infty \varphi(\varepsilon^{-1}A^\sim(\alpha)B^\sim(\alpha))\,d\alpha$$
$$\le 2C\left(\int_0^\infty f(C\delta\varepsilon^{-1}\cdot A^\sim(\alpha))\,d\alpha + \int_0^\infty g(\delta^{-1}B^\sim(\alpha))\,d\alpha\right)$$
$$< 2C(\beta+\delta) < \varepsilon.$$

Thus (b) is proved.

Since the proof of (b) and condition $(*)$ are symmetric with respect to f and g, we also have the following:

(b') $\forall B \in L_g \ \forall \varepsilon > 0 \ \exists \delta > 0$: $(A \in L_f, \|A\|_f < \delta \Rightarrow \|AB\|_\varphi < \varepsilon)$.

By applying the triangle inequality for $\|\cdot\|_\varphi$ to the identity

$$A_n B_n - AB = (A_n - A)(B_n - B) + A(B_n - B) + (A_n - A)\cdot B$$

$(A, A_n \in L_f, \ B, B_n \in L_g, \ n \in \mathbb{N})$ and using (a), (b), and (b'), we obtain (c).

(ii) As in [1], we set $\varphi^{-1} = f^{-1}\cdot g^{-1}$ and let $A \in L_f, \|A\|_f < 1$. Then $B = \varphi(|A|) \in L_1$, $B_1 = f^{-1}(B) \in L_f$, $B_2 = g^{-1}(B) \in L_g$, and $|A| = B_1\cdot B_2 \in L_f\cdot L_g$.

COROLLARY 2.10. *If*

$$(f^{-1})^2 \prec f^{-1}, \qquad (*_1)$$

then $\langle L_f, \|\cdot\|_f\rangle$ is a topological $*$-algebra. All the functions f in Φ_1 satisfy $(*_1)$ and in the case $\tau(1) < \infty$, $L_f = \mathscr{M}$, convergence in $\|\cdot\|_f$ is the same as convergence in $\|\cdot\|_\infty$. The function $g(\lambda) = \lambda/(1+\lambda)$ in Δ_2 also satisfies $(*_1)$; if $\tau(1) < \infty$, then $L_g = \mathscr{K}$ and $\|\cdot\|_g$ define the topology of convergence in the measure t_τ.

PROPOSITION 2.11. *If $\mathscr{M}$ is a von Neumann algebra with a faithful normal semifinite trace τ, $\mathscr{M}$ is not of type I, and $\tau(E_P) = \infty$, then the converses of (i) of Proposition 2.8 and of Lemma 2.9 hold.*

PROOF. If the relation $f \prec g$ is false, then there exists a sequence (λ_n) such that $\lambda_n \in (0, n^{-1}r)$ and $f(\lambda_n) > ng(n\lambda_n)$ $(n \in \mathbb{N})$. Let (α_n) be a sequence of positive numbers such that

$$\sum_{n\ge 1}\alpha_n g(n\lambda_n) < \infty, \qquad \sum_{n\ge 1}\alpha_n n g(n\lambda_n) = \infty.$$

We set $A = \sum_{n\ge 1} n\lambda_n P_n$, where (P_n) is a pairwise orthogonal sequence in $\mathscr{M}^P$ and $\tau(P_n) = \alpha_n$ $(n \in \mathbb{N})$. Let $N_\lambda = \{n \in \mathbb{N}\mid n\lambda_n > \lambda\}$ for $\lambda \in (0, r)$. Then

$$\tau(1 - E_\lambda^{|A|}) = \sum_{n \in N_\lambda}\alpha_n \le \sum_{n\ge 1}\alpha_n\frac{g(n\lambda_n)}{g(\lambda)} < \infty$$

and $A \in \mathscr{K}$. It is clear that $A \in L_g$; but for any $l > 0$ we have

$$\tau(f(l \cdot |A|)) = \sum_{n \geq 1} f(l n \lambda_n) \tau(P_n) \geq \sum_{n \geq l^{-1}} n g(n \lambda_n) = \infty,$$

therefore $A \notin L_f$.

For the proof of the converse of Lemma 2.9 we set $\psi^{-1} = f^{-1} \cdot g^{-1}$. If $L_\varphi = L_f \cdot L_g$, then $L_\psi = L_\varphi \cdot L_g = L_\varphi$ and $\varphi^{-1} \asymp \psi^{-1} = f^{-1} \cdot g^{-1}$. If $L_f \cdot L_g \subset L_\varphi$, then $L_f \cdot L_g = L_\psi \subset L_\varphi$ and, according to what has been proved above, $\varphi \prec \psi$, that is, $\psi^{-1} \prec \varphi^{-1}$ and $f^{-1} \cdot g^{-1} \prec \varphi^{-1}$.

§3. Local boundedness of the spaces $\langle L_f, \|\cdot\|_f \rangle$

THEOREM 3.1. *Let* $f^\dagger(\lambda) = \inf\{t > 0 | 2f(t\lambda) \geq f(\lambda)\}$, $0 < \lambda < r$.

(i) *If* $\inf_{0 < \lambda < r} f^\dagger(\lambda) > 0$, *then* $\langle L_f, \|\cdot\|_f \rangle$ *is locally bounded.*

(ii) *If* $\mathscr{M}$ *is a von Neumann algebra with a faithful normal semifinite trace* τ, $\mathscr{M}$ *is not of type* I, $f \in \Phi_\infty$, *and* $\liminf_{\lambda \to \infty} f^\dagger(\lambda) = 0$, *then* $\langle L_f, \|\cdot\|_f \rangle$ *is not locally bounded.*

PROOF. (i) Let $\alpha = \inf_{0 < \lambda < r} f^\dagger(\lambda)$ and $\varphi(n) = \max\{\kappa \in \mathbb{N} | \alpha^\kappa \geq 2^{\kappa-n}\}$ for $n \geq 1 - \log_2 \alpha$. It is clear that $\varphi(n) \to \infty$ as $n \to \infty$. We prove that the open set $\{A \in L_f | \|A\|_f < 1\}$ is bounded:

$$\|\alpha^n A\|_f = \inf\{\varepsilon > 0 | \tau(f(\alpha^n \cdot \varepsilon^{-1} \cdot |A|)) < \varepsilon\}$$
$$\leq \inf\{\varepsilon > 0 | \tau(f(\alpha^{n-1} \varepsilon^{-1} |A|)) < 2\varepsilon\}$$
$$\leq \inf\{\varepsilon > 0 | \tau(f(\alpha^{\varphi(n)} \cdot \varepsilon^{-1} |A|)) < 2^{n-\varphi(n)} \varepsilon\} \leq \alpha^{\varphi(n)}.$$

(ii) There exists a sequence (λ_n) such that $\lambda_n \to \infty$ and $a_n = f^\dagger(\lambda_n) \to 0$ as $n \to \infty$. Let $\varepsilon > 0$ be sufficiently small so that there exist $P_n \in \mathscr{M}^\mathrm{P}$ such that $\tau(P_n) = \varepsilon/f(\lambda_n)$ $(n \in \mathbb{N})$. We set $A_n = \varepsilon \lambda_n P_n$ $(n \in \mathbb{N})$. Then $\|A_n\|_f = \inf\{\delta > 0 | f(\delta^{-1} \varepsilon \lambda_n) \tau(P_n) < \delta\} = \varepsilon$ $(n \in \mathbb{N})$. We have

$$\tau(f(a_n \varepsilon^{-1} |A_n|)) = f(f^\dagger(\lambda_n) \cdot \lambda_n) \tau(P_n)$$
$$= 2^{-1} f(\lambda_n) \cdot \tau(P_n) = \frac{\varepsilon}{2} \quad (n \in \mathbb{N})$$

and by (i) of Lemma 2.5, $(\|a_n \varepsilon^{-1} A_n\|_f) \notin \Theta$.

COROLLARY 3.2. *Let* $\mathscr{M}$ *be a von Neumann algebra with a faithful normal semifinite trace* τ, *with* $\mathscr{M}$ *not of type* I, $\tau(E_\mathrm{p}) = \infty$, $f \in \Phi_\infty$, *and* L_f *a* $*$-*algebra. Then* $\langle L_f, \|\cdot\|_f \rangle$ *is not a locally bounded topological* $*$-*algebra.*

PROOF. It follows from Corollary 2.10 and Proposition 2.11 that $(f^{-1})^2 \prec f^{-1}$, therefore $\langle L_f, \|\cdot\|_f \rangle$ is a topological $*$-algebra. It is not difficult to verify the inequality $f(\lambda^2) < Cf(C\lambda)$ for some $C > 1$, where $0 < \lambda < \infty$. If $\liminf_{\lambda \to \infty} f^\dagger(\lambda) > 0$ then $f^\dagger(\lambda) \geq \alpha > 0$ for all $\lambda \geq \beta$ for some $\alpha, \beta >$

0. Let $n \in \mathbb{N}$ be such that $2^n > C$ and let $\lambda > \max\{C\alpha^{-n}, \beta\}$. Since $2f(\alpha\lambda) < f(\lambda)$ for all $\lambda \geq \beta$, we have:

$$f(\lambda)^2 > f(C\alpha^{-n}\lambda) > 2f(\alpha^{1-n}C\lambda) > \cdots > 2^n f(C\lambda) > Cf(C\lambda).$$

The contradiction so obtained completes the proof.

§4. The space L_f^* of $\|\cdot\|_f$-continuous linear functionals on L_f

THEOREM 4.1. *If $f \in \Phi_\infty$ and $\liminf_{\lambda \to \infty} f(\lambda)/\lambda > 0$, then $L_f^* \neq \{0\}$.*

PROOF. There exist $\beta, t > 0$ such that $f(\lambda) \geq \beta\lambda$ for $\lambda > t$. Let $A_1, \ldots, A_n \in L_f$, $\|A_i\|_f < 1$, and let $A_i = V_i \cdot |A_i|$ be the polar decompositions ($i = 1, \ldots, n$). It is easy to show that $A_i' = V_i' \cdot |A_i'|$, where $V_i' \equiv V_i(\mathbb{1} - E_t^{|A_i|})$ and $|A_i'| = |A_i| \cdot (\mathbb{1} - E_t^{|A_i|})$ is the polar decomposition of the operator $A_i' = A_i(\mathbb{1} - E_t^{|A_i|})$. Then

$$|A_i'| = \int_{t^+}^\infty \lambda \, dE_\lambda^{|A_i|}.$$

In similar fashion it can be established that

$$|A_i''| = \int_0^t \lambda \, dE_\lambda^{|A_i|} \quad \text{for } A_i'' = A_i - A_i'.$$

We set $A_0' = \frac{1}{n} \sum_{i=1}^n A_i'$, $A_0'' = \frac{1}{n} \sum_{i=1}^n A_i''$, and let $P \in \mathscr{M}^{\mathrm{P}}$, $\tau(P) = \alpha \in (0, \infty)$. Since $\tau(|A_i''P|) < t\alpha$, it follows that $\tau(|A_0''P|) \leq t\alpha$ as well. We then have

$$\beta\tau(|A_i'P|) \leq \beta \int_{t^+}^\infty \lambda\tau(dE_\lambda^{|A_i|}) \leq \int_{t^+}^\infty f(\lambda)\tau(dE_\lambda^{|A_i|}) \leq \tau(f(|A_i|)) < 1,$$

therefore $\tau(|A_i'P|) < \beta^{-1}$ and $\tau(|A_0'P|) < \beta^{-1}$.

It is now clear that the operator $(t + 1/\alpha\beta) \cdot P$ does not belong to the convex hull of the set $\{A \in L_f \mid \|A\|_f < 1\}$.

BIBLIOGRAPHY

1. A. L. Vol′berg and V. A. Tolokonnikov, *Hankel operators and problems of best approximation of unbounded functions*, Zap. Nauchn. Sem. Leningrad. Otdel. Mat. Inst. Steklov. (LOMI) **141** (1985), 5–17.
2. Nelson Dunford and Jacob T. Schwartz, *Linear operators. Part II: Spectral theory. Self adjoint operators in Hilbert space*, Interscience, New York, 1963.
3. I. E. Segal, *Non-commutative extension of integration*, Ann. of Math. (2) **57** (1953), 401–457.
4. S. Rolewicz, *On a certain class of linear metric spaces*, Bull. Acad. Polon. Sci. Cl. III **5** (1957), 471–473.
5. F. J. Yeadon, *Non-commutative L_p-spaces*, Math. Proc. Cambridge Philos. Soc. **77** (1975), 91–102.

Translated by G. G. GOULD

Generalized Resolvents of Nondensely Defined Bounded Symmetric Operators

A. V. SHTRAUS

We shall consider a symmetric nonselfadjoint operator A in a Hilbert space $\mathfrak{H}$. Unless explicitly stated otherwise, we assume this operator to be bounded and defined on a subspace $\mathscr{D}(A) = \mathfrak{M} \neq \mathfrak{H}$. We are going to describe the collection all of generalized resolvents of the operator A.

Let $\mathfrak{N} = \mathfrak{H} \ominus \mathfrak{M}$ and denote by P and Q the orthogonal projections onto $\mathfrak{M}$ and $\mathfrak{N}$, respectively. Let $T = PA$, $\rho(T)$ be the resolvent set of the operator T, when the latter is regarded as a selfadjoint operator on $\mathfrak{M}$, $V = QA$, and
$$\mathfrak{M}_\lambda = \mathscr{R}(A - \lambda I) = (A - \lambda I)\mathfrak{M}$$
for every complex λ.

THEOREM 1. *The equality*
$$\mathfrak{H} = \mathfrak{M}_\lambda + \mathfrak{N} \tag{1}$$
holds if and only if $\lambda \in \rho(T)$. *In that case* $\mathfrak{M}_\lambda$ *is a subspace* $\mathfrak{M}_\lambda \cap \mathfrak{N} = \{0\}$, *and so the linear sum on the right-hand side of* (1) *is direct, i.e.,* $\mathfrak{H} = \mathfrak{M}_\lambda \dotplus \mathfrak{N}$. *For every* $\lambda \in \rho(T)$, *the operators*
$$P(\lambda) = (A - \lambda I)(T - \lambda I)^{-1} P \tag{2}$$
and
$$Q(\lambda) = Q - V(T - \lambda I)^{-1} P \tag{3}$$
are the projections onto $\mathfrak{M}_\lambda$ *and* $\mathfrak{N}$, *respectively, corresponding to the decomposition* (1).

PROOF.[1] Straightforward verification reveals that, for every $\lambda \in \rho(T)$, $P^2(\lambda) = P(\lambda)$, $\operatorname{Ker} P(\lambda) = \mathfrak{N}$, and $P(\lambda)\mathfrak{H} = \mathfrak{M}_\lambda$. This implies the direct sum

---

1991 *Mathematics Subject Classification.* Primary 47B25; Secondary 47A20.

Translation of Functional Analysis. Spectral theory, Interinstitutional collection of scientific works, Ul′yanovsk. Gos. Ped. Inst., Ul′yanovsk, 1987, pp. 187–196.

[1] This theorem strengthens one of the results of the paper [1] (Theorem 1).

decomposition (1) and shows that $P(\lambda)$ is the operator of projection onto the subspace $\mathfrak{M}_\lambda$ parallel to the subspace $\mathfrak{N}$. Since $P(\lambda) + Q(\lambda) = I$, it is clear that $Q(\lambda)$ is the projection onto $\mathfrak{N}$ parallel to the subspace $\mathfrak{M}_\lambda$.

Conversely, assume the equality (1). Then $\mathfrak{M} = P\mathfrak{H} = P\mathfrak{M}_\lambda = P(A - \lambda I)\mathfrak{M} = (T - \lambda I)\mathfrak{M}$. Since T is a selfadjoint operator on $\mathfrak{M}$, it follows that $\lambda \in \rho(T)$.

COROLLARY 2. *For arbitrary $\lambda \in \rho(T)$ the subspaces $\mathfrak{M}$ and $\mathfrak{N}_\lambda = \mathfrak{H} \ominus \mathfrak{M}_\lambda$ are linearly independent, $\mathfrak{H} = \mathfrak{M} \dotplus \mathfrak{N}_\lambda$, and the projection onto $\mathfrak{N}_\lambda$ parallel to the subspace $\mathfrak{M}$ is given by the formula*

$$Q^*(\lambda) = (I - (T - \bar{\lambda}I)^{-1}V^*)Q. \qquad (4)$$

Here V^* is the operator from $\mathfrak{N}$ into $\mathfrak{M}$ adjoint to the operator V when we regard the latter as an operator from $\mathfrak{M}$ into $\mathfrak{N}$.

Let us recall the definition of generalized resolvent. Let A be a closed, symmetric, not necessarily bounded operator on a Hilbert space $\mathfrak{H}$, $\widetilde{A}$ its selfadjoint extension, possibly going out into a large Hilbert space $\widetilde{\mathfrak{H}} \supset \mathfrak{H}$, and $\widetilde{P}_\mathfrak{H}$ the orthogonal projection of $\widetilde{\mathfrak{H}}$ onto $\mathfrak{H}$. The operator-valued function

$$\lambda \mapsto R_\lambda = \widetilde{P}_\mathfrak{H}(\widetilde{A} - \lambda I)^{-1}|_\mathfrak{H}$$

of the nonreal variable λ is called the generalized resolvent of the operator A, generated by the extension $\widetilde{A}$.

A linear operator B, acting on a Hilbert space $\mathscr{L}$, will be called accumulative* (dissipative) when $\operatorname{Im}(Bf, f) \leq 0$ (≥ 0), respectively, for arbitrary $f \in \mathscr{D}(B)$. Denote by Π_+ (Π_-) the upper (lower) half-plane in the complex plane.

Let us agree to say that an operator-valued function $\lambda \mapsto B(\lambda)$ of the nonreal variable λ belongs to the class $\mathscr{K}_\mathscr{L}$ when the following conditions are satisfied.

1) For every $\lambda \in \Pi_+$ (Π_-), $B(\lambda)$ is a maximal closed accumulative (dissipative) operator in $\mathscr{L}$.

2) The operator-valued function $\lambda \mapsto B(\lambda)$ is holomorphic in the half-plane Π_+ (Π_-) in the generalized sense. This means that the operator-valued functions $\lambda \mapsto (B(\lambda) - iI)^{-1}$ and $\lambda \mapsto (B(\lambda) + iI)^{-1}$ are boundedly holomorphic (i.e., holomorphic in the usual sense) in the half-planes Π_+ and Π_-, respectively (consult [2], pp. 365–366 on questions of terminology).

3) For any nonreal λ, $B^*(\lambda) = B(\bar{\lambda})$.

Note that every function of class $\mathscr{K}_\mathscr{L}$ is a selfadjoint holomorphic family of operators in the terminology of [2], pp. 385–387; the class of such families, however, is larger than the class $\mathscr{K}_\mathscr{L}$.

A function $\lambda \mapsto B(\lambda)$ of class $\mathscr{K}_\mathscr{L}$ will be called admissible if

$$\lambda(B(\lambda) - \lambda I)^{-1} \to -I$$

---

*Translator's note. The term *accretive* seems to be used more frequently.

in the sense of convergence, i.e., if $B(\lambda)/\lambda \to 0$ in the sense of generalized strong convergence (this kind of convergence is discussed in [2], p. 429) whenever $\lambda \to \infty$ in the angles $\varepsilon < |\arg \lambda| < \pi - \varepsilon$.

The following two theorems are just reformulations of some of the results from [3] and [4].

THEOREM 3. *The collection of all generalized resolvents of a closed symmetric (not necessarily bounded) operator A in $\mathfrak{H}$ can be described by the formula*
$$R_\lambda = (B(\lambda) - \lambda I)^{-1} \quad (\operatorname{Im} \lambda \neq 0),$$
where $\lambda \mapsto B(\lambda)$ is an arbitrary admissible function of class $\mathscr{K}_\mathfrak{H}$ whose values are extensions of the operator A.

THEOREM 4. *Let $\lambda \mapsto B(\lambda)$ be a function of class $\mathscr{K}_\mathfrak{H}$ whose values are extensions of a closed symmetric (not necessarily bounded) operator A on $\mathfrak{H}$. For this function to be admissible it suffices (and is necessary) that*
$$\lim_{\substack{\lambda \to \infty \\ \varepsilon < \arg \lambda < \pi - \varepsilon}} (\lambda(B(\lambda) - \lambda I)^{-1} \varphi, \varphi) = -(\varphi, \varphi)$$
for every $\varphi \in \mathfrak{H} \ominus \overline{\mathscr{D}(A)}$. In particular, in case the operator A is densely defined, every function $\lambda \mapsto B(\lambda) \supset A$ of class $\mathscr{K}_\mathfrak{H}$ is admissible.

REMARK 5. If $\lambda \mapsto B(\lambda)$ is a function of class $\mathscr{K}_\mathfrak{H}$ and $B(\lambda_0) \supset A$ for some nonreal λ_0, then $B(\lambda) \supset A$ for every nonreal λ. To prove this, say, for $\lambda_0 \in \Pi_+$, just consider the operator-valued function $\lambda \mapsto (B(\lambda) + iI) \cdot (B(\lambda) - iI)^{-1}$, whose values in the half-plane Π_+ are contractions, note that the operator $(A + iI)(A - iI)^{-1}$ is an isometry, and apply the generalization of the maximum modulus principle for holomorphic functions with values in $\mathfrak{H}$.

Now let us take up the case when A is a bounded operator defined on a subspace $\mathfrak{M} \neq \mathfrak{H}$, $\mathfrak{N} = \mathfrak{H} \ominus \mathfrak{M}$, and P, Q, T, V are the operators introduced above. We are interested in those extensions B of the operator A that are closed, have domain of definition dense in $\mathfrak{H}$ (it can differ from $\mathfrak{H}$ only if $\mathfrak{N}$ is infinite-dimensional), and satisfy the condition $B^* \supset A$. In this case $QB^*|_\mathfrak{M} = V$, whence $PB|_{\mathfrak{N} \cap \mathscr{D}(B)} \subset V^*$. It is convenient to write such an operator B as a block matrix
$$B = \begin{pmatrix} T & V^* \\ V & W \end{pmatrix}, \tag{5}$$
where W is a closed, densely defined operator in $\mathfrak{N}$; then $\mathscr{D}(B) = \mathfrak{M} \oplus \mathscr{D}(W)$ and, for every $g \in \mathscr{D}(W)$, $QBg = Wg$. Set
$$X(\lambda) = \lambda I + V(T - \lambda I)^{-1} V^* \tag{6}$$
for any $\lambda \in \rho(T)$. In accordance with [5], we shall call the operator-valued function $\lambda \mapsto X(\lambda)$ the spectral complement of A.

LEMMA 6. *Let B be an operator of the form* (5), *and let* $\lambda \in \rho(T)$. *Then* $\lambda \in \rho(B)$ *if and only if the operator* $W - \dot{X}(\lambda)$ *possesses a bounded inverse, defined on the whole of* $\mathfrak{N}$; *in this case*

$$(B - \lambda I)^{-1} = (T - \lambda I)^{-1}P + Q^*(\bar{\lambda})(W - X(\lambda))^{-1}Q(\lambda). \tag{7}$$

PROOF.[2] For $h \in \mathfrak{M}_\lambda$ the equation

$$(B - \lambda I)x = h \tag{8}$$

has an obvious solution $x = (A - \lambda I)^{-1}h$; the operator $(A - \lambda I)^{-1}$ exists and is bounded since $\lambda \in \rho(T)$.

Let us make clear under what conditions the equation (8) is uniquely solvable for arbitrary $h \in \mathfrak{H}$. Setting $h_1 = Ph$, $h_2 = Qh$, $x_1 = Px$, and $x_2 = Qx$, we arrive at the system

$$(T - \lambda I)x_1 + V^*x_2 = h_1, \qquad Vx_1 + (W - \lambda I)x_2 = h_2.$$

For $h = 0$ this reduces to

$$x_1 = -(T - \lambda I)^{-1}V^*x_2, \qquad (W - X(\lambda))x_2 = 0. \tag{9}$$

Consequently, equation (8) with $h = 0$ has only the zero solution if and only if

$$\text{Ker}(W - X(\lambda)) = \{0\}. \tag{10}$$

If $h \in \mathfrak{N}$, then $h_1 = 0$, $h_2 = h$ and this produces a system consisting of equations (9) together with $(W - X(\lambda))x_2 = h$. The last equation is uniquely solvable for each $h \in \mathfrak{N}$ if and only if, in addition to (10), the condition $\mathscr{R}(W - X(\lambda)) = \mathfrak{N}$ is fulfilled; the operator W being closed and $X(\lambda)$ bounded, this means that the operator $W - X(\lambda)$ possesses a bounded inverse, defined on all of $\mathfrak{N}$. The formula for the solution of (8) then reads

$$x = x_1 + x_2 = (-(T - \lambda I)^{-1}V^* + I)(W - X(\lambda))^{-1}h,$$

or, by virtue of (4),

$$x = Q^*(\bar{\lambda})(W - X(\lambda))^{-1}h.$$

Taking into account Theorem 1, we conclude that the condition formulated in Lemma 6 is necessary and sufficient for equation (8) to have a unique solution x for each $h \in \mathfrak{H}$, and in that case

$$x = (A - \lambda I)^{-1}P(\lambda)h + Q^*(\bar{\lambda})(W - X(\lambda))^{-1}Q(\lambda)h.$$

In view of (2),

$$(A - \lambda I)^{-1}P(\lambda) = (T - \lambda I)^{-1}P,$$

and formula (7) follows.

---

[2] This lemma completes and generalizes Proposition 3 of [5], which concerns bounded extensions of A.

COROLLARY 7. *Under the hypothesis of Lemma 6,*

$$Q(B - \lambda I)^{-1}|_{\mathfrak{N}} = (W - X(\lambda))^{-1}$$

for every $\lambda \in \rho(B)$.

Indeed, this follows immediately from (7), since $QQ^*(\bar{\lambda}) = Q$ owing to (4).

THEOREM 8. *The collection of all generalized resolvents of* A *can be described by the formula*

$$R_\lambda = \left(\begin{pmatrix} T & V^* \\ V & W(\lambda) \end{pmatrix} - \lambda I \right)^{-1} \qquad (\operatorname{Im} \lambda \neq 0),$$

where $\lambda \mapsto W(\lambda)$ *stands for an arbitrary admissible function of class* $\mathscr{H}_{\mathfrak{N}}$.

PROOF. In view of Theorem 3 it suffices to check that an operator-valued function $\lambda \mapsto B(\lambda)$ is admissible, of class $\mathscr{H}_{\mathfrak{H}}$, and such that $B(\lambda) \supset A$ for each nonreal λ if and only if it has the form

$$B(\lambda) = \begin{pmatrix} T & V^* \\ V & W(\lambda) \end{pmatrix} \qquad (11)$$

for some admissible function $\lambda \mapsto W(\lambda)$ of class $\mathscr{H}_{\mathfrak{N}}$. This is quite obvious when $\mathfrak{N}$ is finite-dimensional, or, more generally, when the values of the functions $\lambda \mapsto B(\lambda)$ and $\lambda \mapsto W(\lambda)$ are bounded operators. In the general case, note first of all that an operator $B(\lambda)$ of the form (11) is maximal closed accumulative (dissipative) in $\mathfrak{H}$, for some $\lambda \in \Pi_+$ (Π_-), if and only if the operator $W(\lambda)$ in $\mathfrak{N}$ enjoys that property. Likewise, the equality $B^*(\lambda) = B(\bar{\lambda})$ is equivalent to $W^*(\lambda) = W(\bar{\lambda})$ under the given conditions. Consequently, according to Lemma 6 and Corollary 7,

$$(B(\lambda) - iI)^{-1} = (T - iI)^{-1} P + Q^*(-i)(W(\lambda) - X(i))^{-1} Q(i), \qquad (12)$$

$$Q(B(\lambda) - iI)^{-1}|_{\mathfrak{N}} = (W(\lambda) - X(i))^{-1} \qquad (13)$$

for every $\lambda \in \Pi_+$,

$$(B(\lambda) + iI)^{-1} = (T + iI)^{-1} P + Q^*(i)(W(\lambda) - X(-i))^{-1} Q(-i), \qquad (14)$$

$$Q(B(\lambda) + iI)^{-1}|_{\mathfrak{N}} = (W(\lambda) - X(-i))^{-1} \qquad (15)$$

for every $\lambda \in \Pi_-$, and

$$(B(\lambda) - \lambda I)^{-1} = (T - \lambda I)^{-1} P + Q^*(\bar{\lambda})(W(\lambda) - X(\lambda))^{-1} Q(\lambda), \qquad (16)$$

$$Q(B(\lambda) - \lambda I)^{-1}|_{\mathfrak{N}} = (W(\lambda) - X(\lambda))^{-1} \qquad (17)$$

for every nonreal λ. The equalities (12) and (13) ((14) and (15)) imply that the function $\lambda \mapsto B(\lambda)$ is holomorphic in the half-plane Π_+ (Π_-) if and only if the function $\lambda \mapsto W(\lambda)$ is holomorphic in that half-plane. Finally, formulas (16), (17), (3), (4), and (6) imply that the function $\lambda \mapsto B(\lambda)$ is admissible of class $\mathscr{K}_{\mathfrak{H}}$ if and only if the corresponding function $\lambda \mapsto W(\lambda)$ is admissible of class $\mathscr{K}_{\mathfrak{N}}$; just observe that, owing to (6),

$$(W(\lambda) - X(\lambda))^{-1} - (W(\lambda) - \lambda I)^{-1}$$
$$= (W(\lambda) - X(\lambda))^{-1} V (T - \lambda I)^{-1} V^* (W(\lambda) - \lambda I)^{-1}$$

and the norms of the three operators $(W(\lambda) - X(\lambda))^{-1}$, $(W(\lambda) - \lambda I)^{-1}$, and $(T - \lambda I)^{-1}$ do not exceed $1/|\operatorname{Im} \lambda|$.

In the course of the proof of Theorem 8, we have established formulas (16) and (17), thereby proving the following theorem.

THEOREM 9. *The formula*

$$R_\lambda = (T - \lambda I)^{-1} P + Q^*(\bar{\lambda})(W(\lambda) - X(\lambda))^{-1} Q(\lambda) \qquad (\operatorname{Im} \lambda \neq 0)$$

establishes a bijective correspondence between the collection of all admissible functions $\lambda \mapsto W(\lambda)$ *of class* $\mathscr{K}_{\mathfrak{N}}$ *and the collection of all generalized resolvents of the operator* A. *Also*

$$Q R_\lambda|_{\mathfrak{N}} = (W(\lambda) - X(\lambda))^{-1} \qquad (\operatorname{Im} \lambda \neq 0).$$

In case A is a contraction, the collection of all generalized resolvents of A described above contains the class of those generalized resolvents that correspond to contractive selfadjoint extensions $\widetilde{A}$ of A, possibly going out of the space $\mathfrak{H}$. This class of generalized resolvents was studied by M. G. Kreĭn and I. E. Ovcharenko [6]. Another case which deserves to be mentioned occurs when A is bounded, positive, nondensely defined, and admits a bounded or unbounded positive selfadjoint extension. A criterion for existence of such an extension was first established by T. Ando and K. Nishio [7]. Another criterion, in terms of the spectral complement $X(\lambda)$, was obtained in [5]. We also mention the article of M. G. Kreĭn and I. E. Ovcharenko [8] where generalized resolvents of densely defined positive symmetric operators are discussed and which also provides bibliographic references concerning that topic. The applications of Theorem 9 to the particular cases just mentioned (i.e., when A is a contraction or a bounded positive operator) will be published elsewhere.

BIBLIOGRAPHY

1. A. V. Shtraus, *Extensions of a bounded symmetric operator with preservation of the bound*, Functional Analysis, No. 6, Ul′yanovsk. Gos. Ped. Inst., Ul′yanovsk, 1976, pp. 165–173.
2. T. Kato, *Perturbation theory for linear operators*, Springer-Verlag, Berlin and New York, 1984.
3. A. V. Shtraus, *Generalized resolvents of symmetric operators*, Izv. Akad. Nauk SSSR Ser. Mat. **18** (1954), 51–86. (Russian)
4. _____, *Extensions and generalized resolvents of a non-densely defined symmetric operator*, Izv. Akad. Nauk SSSR Ser. Mat. **34** (1970), 175–202; English transl. in Math. USSR-Izv. **4** (1970).
5. _____, *On the theory of extremal extensions of a bounded positive operator*, Functional Analysis, No. 18, Ul′yanovsk. Gos. Ped. Inst., Ul′yanovsk, 1982, pp. 115–126.

6. M. G. Kreĭn and I. E. Ovcharenko, *On the theory of generalized resolvents of non-densely defined Hermitian contractions*, Dokl. Akad. Nauk Ukrain. SSR **10** (1976), 881–885. (Russian)
7. T. Ando and K. Nishio, *Positive selfadjoint extensions of positive symmetric operators*, Tôhoku Math. J. (2) **22** (1970), 65–75.
8. M. G. Kreĭn and I. E. Ovcharenko, *The generalized resolvents and resolvent matrices of positive Hermitian operators*, Dokl. Akad. Nauk SSSR **231** (1976), 1063–1066; English transl. in Soviet Math. Dokl. **17** (1976).

Translated by M. ENGLIŠ

Recent Titles in This Series

(Continued from the front of this publication)

114 **M. Š. Birman and M. Z. Solomjak,** Quantitative Analysis in Sobolev Imbedding Theorems and Applications to Spectral Theory
113 **A. F. Lavrik,** Twelve Papers in Logic and Algebra
112 **D. A. Gudkov and G. A. Utkin,** Nine Papers on Hilbert's 16th Problem
111 **V. M. Adamjan, et al.,** Nine Papers on Analysis
110 **M. S. Budjanu, et al.,** Nine Papers on Analysis
109 **D. V. Anosov, et al.,** Twenty Lectures Delivered at the International Congress of Mathematicians in Vancouver, 1974
108 **Ja. L. Geronimus and Gábor Szegő,** Two Papers on Special Functions
107 **A. P. Mišina and L. A. Skornjakov,** Abelian Groups and Modules
106 **M. Ja. Antonovskiĭ, V. G. Boltjanskiĭ, and T. A. Sarymsakov,** Topological Semifields and Their Applications to General Topology
105 **R. A. Aleksandrjan, et al.,** Partial Differential Equations, Proceedings of a Symposium Dedicated to Academician S. L. Sobolev
104 **L. V. Ahlfors, et al.,** Some Problems on Mathematics and Mechanics, On the Occasion of the Seventieth Birthday of Academician M. A. Lavrent'ev
103 **M. S. Brodskiĭ, et al.,** Nine Papers in Analysis
102 **M. S. Budjanu, et al.,** Ten Papers in Analysis
101 **B. M. Levitan, V. A. Marčenko, and B. L. Roždestvenskiĭ,** Six Papers in Analysis
100 **G. S. Ceĭtin, et al.,** Fourteen Papers on Logic, Geometry, Topology and Algebra
99 **G. S. Ceĭtin, et al.,** Five Papers on Logic and Foundations
98 **G. S. Ceĭtin, et al.,** Five Papers on Logic and Foundations
97 **B. M. Budak, et al.,** Eleven Papers on Logic, Algebra, Analysis and Topology
96 **N. D. Filippov, et al.,** Ten Papers on Algebra and Functional Analysis
95 **V. M. Adamjan, et al.,** Eleven Papers in Analysis
94 **V. A. Baranskiĭ, et al.,** Sixteen Papers on Logic and Algebra
93 **Ju. M. Berezanskiĭ, et al.,** Nine Papers on Functional Analysis
92 **A. M. Ančikov, et al.,** Seventeen Papers on Topology and Differential Geometry
91 **L. I. Barklon, et al.,** Eighteen Papers on Analysis and Quantum Mechanics
90 **Z. S. Agranovič, et al.,** Thirteen Papers on Functional Analysis
89 **V. M. Alekseev, et al.,** Thirteen Papers on Differential Equations
88 **I. I. Eremin, et al.,** Twelve Papers on Real and Complex Function Theory
87 **M. A. Aĭzerman, et al.,** Sixteen Papers on Differential and Difference Equations, Functional Analysis, Games and Control
86 **N. I. Ahiezer, et al.,** Fifteen Papers on Real and Complex Functions, Series, Differential and Integral Equations
85 **V. T. Fomenko, et al.,** Twelve Papers on Functional Analysis and Geometry
84 **S. N. Černikov, et al.,** Twelve Papers on Algebra, Algebraic Geometry and Topology
83 **I. S. Aršon, et al.,** Eighteen Papers on Logic and Theory of Functions
82 **A. P. Birjukov, et al.,** Sixteen Papers on Number Theory and Algebra
81 **K. K. Golovkin, V. P. Il'in, and V. A. Solonnikov,** Four Papers on Functions of Real Variables
80 **V. S. Azarin, et al.,** Thirteen Papers on Functions of Real and Complex Variables
79 **V. I. Arnol'd, et al.,** Thirteen Papers on Functional Analysis and Differential Equations
78 **A. V. Arhangel'skiĭ, et al.,** Eleven Papers on Topology

(See the AMS catalog for earlier titles)